Diversions in Modern Mathematics

Diversions in Modern Mathematics

Barry Lewis

Heinemann Educational Books

Heinemann Educational Books Ltd
22 Bedford Square, London WC 1 B 3 HH

LONDON EDINBURGH MELBOURNE AUCKLAND
HONG KONG SINGAPORE KUALA LUMPUR NEW DELHI
IBADAN NAIROBI JOHANNESBURG
EXETER (NH) KINGSTON PORT OF SPAIN

ISBN 0 435 51600 0

First published 1981

Photoset by Thomson Press (India) Limited, New Delhi and
Printed and bound in Great Britain by Spottiswoode Ballantyne Ltd.,
Colchester and London

Foreword

In these days of shortages of staff, there is little time for individual mathematics schoolmasters, such as Mr. B. Lewis, to take up the challenge of trying to write down something of what they have found from their experience to be acceptable and useful in the classroom, unless they are lucky enough to find themselves supported in project or research work, supported privately or publicly from outside the school.

One can therefore only wish well any such schoolmaster author who, of his own initiative, is willing to give us a glimpse of his work in class, trying to achieve the difficult task of translating lesson notes into published work available inevitably for all to criticise.

Would that more of the profession had the time and energy to do what was so often taken for granted in the past!

W. H. Cockcroft

Preface

The work covered in this book was originally designed as a series of lectures to supplement and enrich a modern mathematical sixth form course. The lectures were given at the end of each term, and in so far as I could arrange it, they were to be integrated with, or at least a stimulus to, the work following in the next term.

The course as it now stands has been developed and modified a great deal from its rather humble and faltering start some years ago. The original source material were classics, like *Mathematical Recreations and Essays* by W. W. R. Ball. However, like many other very stimulating books of a similar type, they all suffer one major defect—very often there is a complete absence of proofs, formal or otherwise. For instance, on p. 61 of the revised edition of the above book (Editor: H.S.M. Coxeter) we learn,
'... and a problem, such as that of finding a positive integer x for which,

$$4\,787\,532\,864x^2 + 171\,576\,027\,618\,912x + 10\,553\,711\,710\,905$$

is a square (this problem arises quite *naturally* in seeking to factorize $2^{93} + 1$)'! (my italics).

Fascinating, no doubt, but a little frustrating too. The lectures that I originally gave were similarly received, and I decided that I must supply proofs in my development and problems suitable for further work. Thus I re-wrote the lectures and developed the problems that accompany them.

The course, as I am now presenting it, follows the lines of this book, except that I now give two lectures a term instead of one. It creates a good deal of mathematical interest and an awareness of what constitutes a proof. Additionally, apart from some standard results, it is entirely self-contained.

B.L.
1981

Contents

1 Some mathematics from antiquity

1.1 Magic squares

Man has always been an orderly animal and each step forward in his development has been as the result of a re-ordering of his perceptions of the world. Thus we find that almost as soon as man could count, he started looking for relations amongst his numbers that could not be of any practical use to him. Man was not content merely to count and measure for he sought the subtle patterns behind the practical tools he had created.

Consider Figure 1.1, which comes to us from a copy of a Chinese manuscript dated about 2200 BC.

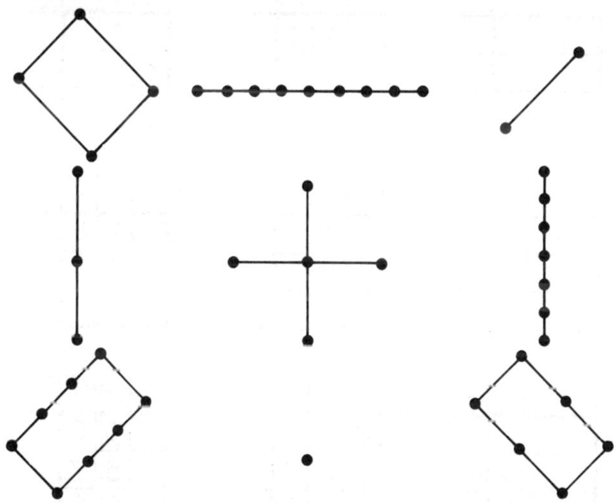

Fig. 1.1 Illustration from ancient Chinese manuscript (2200 BC)

This is a pictorial illustration of what is known as a *magic square.*

Can you see (Figure 1.2) that the sum of the numbers in any row, or any column, or diagonal is always 15? Because such squares appeared to be magic, mystical properties have been associated with them. The theory of magic squares has continued to be of interest, and a great deal is now known about them.

4	9	2
3	5	7
8	1	6

Fig. 1.2 Typical magic square

We shall say that a magic square has order n if it is a square array of n^2 consecutive numbers beginning with 1. Thus the example above is said to have order 3.

Q 1.1 Is it possible to find a magic square of order 2?

A very simple method of constructing a magic square of any odd order was discovered by de la Loubere, when he was the ambassador of Louis

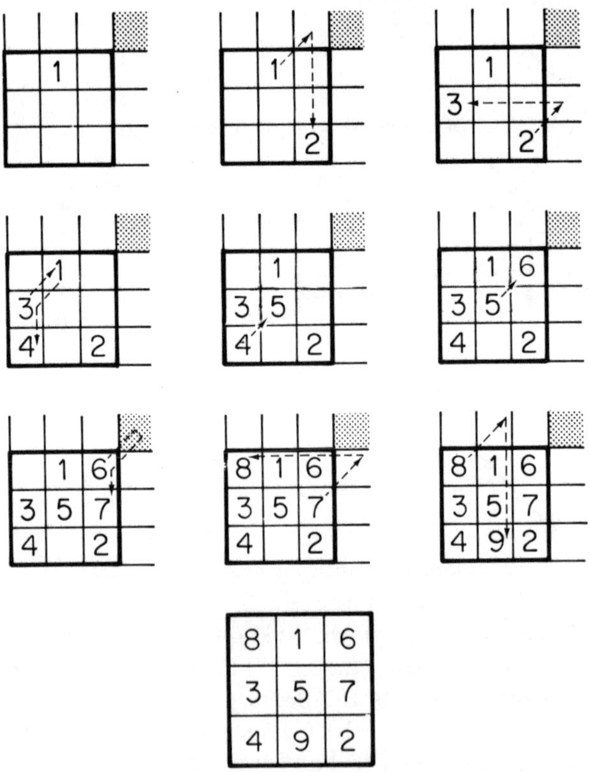

Fig. 1.3 Sequence for construction of a magic square

XIV to the King of Siam. This method had been known to the Siamese for some time, and it seems that it was discovered by chance. We can illustrate the method by constructing a magic square of the third order. Draw a square and divide it into nine *cells*. Border the square with cells along the top and the right edge, and shade the added cell in the top right hand corner. Write 1 in the middle top cell of the original square as shown, and then proceed as follows (see Figure 1.3). The general rule is to proceed upward to the right with the next number except where,

(i) this takes one onto the border, in which case one goes from the border back into the original square by shifting clear across the square, either from top to bottom or from right to left,
(ii) a move terminates in a square already occupied when the move is disallowed and one proceeds immediately below the last written entry. The shaded square is to be regarded as occupied.

Q 1.2 Construct magic squares of the fifth and seventh orders.

There is an easy way to find out the *magic constant* of any magic square. A magic square of the nth order consists of the numbers,

$$1, 2, 3, \ldots, n^2$$

arranged one in each of the n^2 cells. Hence, the sum of all these numbers is,

$$S = 1 + 2 + 3 + \ldots + n^2 - 2 + n^2 - 1 + n^2 \tag{1}$$

We can write this sum S in another way:

$$S = n^2 + n^2 - 1 + n^2 - 2 + \ldots + 3 + 2 + 1 \tag{2}$$

If we add (1) and (2) column by column, we obtain:

$$\begin{aligned} 2S &= (n^2 + 1) + (n^2 - 1 + 2) + (n^2 - 2 + 3) + \ldots + (n^2 + 1) \\ &= (n^2 + 1) + (n^2 + 1) + (n^2 + 1) + \ldots + (n^2 + 1) \\ &= n^2(n^2 + 1) \text{ since there are } n^2 \text{ terms} \end{aligned}$$

Hence, $S = n^2(n^2 + 1)/2$.
So the sum of all the numbers in the magic square is this number S. Let the total of each column be x. Then since there are n columns, the total of all the numbers in the square is nx. Hence,

$$nx = \frac{n^2(n^2 + 1)}{2}$$

$$x = \frac{n^2(n^2 + 1)}{2n} = \frac{n(n^2 + 1)}{2}$$

Since the column total is the same as the row total and the diagonal total, this number x represents the magic constant of the square.

Q 1.3 (i) What is the magic constant of a square of order 3?
(ii) Use this to try and find a magic square of order 3 different from the one just obtained.

There are various ways of constructing larger order magic squares, apart from when they are odd. However, in order to make it clear we begin with a few simple cases. Let us consider the magic square of order 4. This has a magic constant of $4(4^2+1)/2=34$. First of all we begin with the most obvious square (Figure 1.4).

1	2	3	4
5	6	7	8
9	10	11	12
13	14	15	16

Fig. 1.4 First stage of magic square construction

We note that both diagonals have the required sum of 34. However, the sum of the top rows is too small, while that of the lower rows is too large. Similarly, the sum of the first two columns is too small, and that of the last two is too large. This suggests that we should leave the sum of the diagonals as they are by interchanging entries on the diagonals from top to bottom, and simultaneously right to left (Figure 1.5).

16	2	3	13
5	11	10	8
9	7	6	12
4	14	15	1

Fig. 1.5 A half turn of the diagonals

We call such a transformation a half turn of the diagonals. Happily, this is now a magic square of order 4. But does the method always work? Let us try it out on the magic square of order 6. First of all we note that the magic constant of such a square is $6(6^2+1)/2=111$. If we perform a half-turn of the diagonals of the most obvious square of this type, we obtain the square shown in Figure 1.6.

36	2	3	4	5	31
7	29	9	10	26	12
13	14	22	21	17	18
19	20	16	15	23	24
25	11	27	28	8	30
6	32	33	34	35	1

Fig. 1.6 A half turn of the diagonals on square of order 6

If we examine this in detail we find that,

Row	*Sum*	*Difference*
First	81	$111-81=+30$
Second	93	$111-93=+18$
Third	105	$111-105=+6$
Fourth	117	$111-117=-6$
Fifth	129	$111-129=-18$
Sixth	141	$111-141=-30$

In order to obtain a sum of 111 for each row we need to interchange one element between the first and final rows and this element must not lie on the diagonals. Similarly, we need to interchange elements between the second and fifth rows, and between the third and fourth rows. Before making these necessary interchanges, we note in the same way the sum for each column.

Column	*Sum*	*Difference*
First	106	$111-106=+5$
Second	108	$111-108=+3$
Third	110	$111-110=+1$
Fourth	112	$111-112=-1$
Fifth	114	$111-114=-3$
Sixth	116	$111-116=-5$

Clearly, we also need to interchange elements in the columns in exactly the same way. One way of doing this is shown in Figure 1.7 (remember that an element is allowed to move only once, and those on the diagonal not at all—can you explain why?);

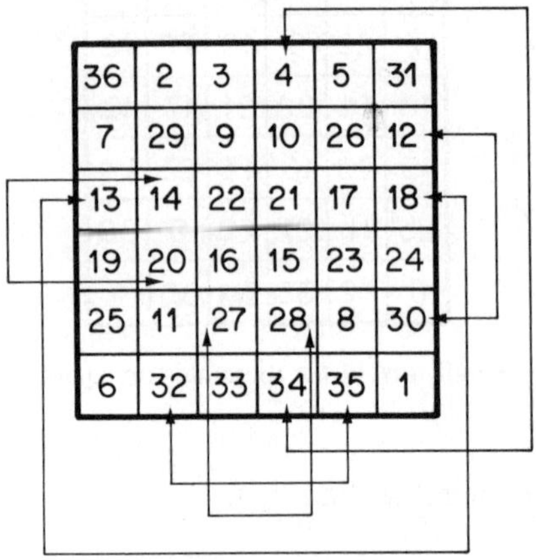

Fig. 1.7 Interchanging elements in the columns shown

This gives the magic square shown in Figure 1.8.

36	2	3	34	5	31
7	29	9	10	26	30
18	20	22	21	17	13
19	14	16	15	23	24
25	11	28	27	8	12
6	35	33	4	32	1

Fig. 1.8 Result of interchanging shown in Figure 1.7

Q 1.4 Find another magic square of order 6 by changing different entries from the ones above.

Magic squares have other properties which are equally pleasing. Claude Brogdan, an American architect, discovered that on most magic squares a line traced from cell to cell in the order of the numbers produces an artistically pleasing pattern.

Some of these patterns are shown in Figure 1.9.

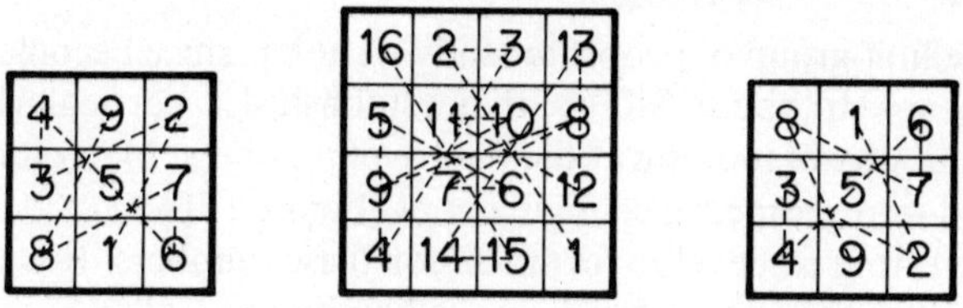

Fig. 1.9 Artistic patterns produced by linking numbers within magic squares

Q 1.5 Draw some 5 × 5 squares and construct on them some patterns like the ones in Figure 1.9. Try to use these patterns to construct magic squares.

It is a surprising fact but although a great deal is known about magic squares, little is known about the actual number that can be constructed. Thus, it has been estimated that there are at least 13 million of order 5. Happy hunting.

Q 1.6 What number must occupy the central cell of a magic square of order 3?

1.2 Figurative numbers

The Greeks, of course, laid many of the foundations on which most of the future developments in mathematics were made. The Pythagoreans were

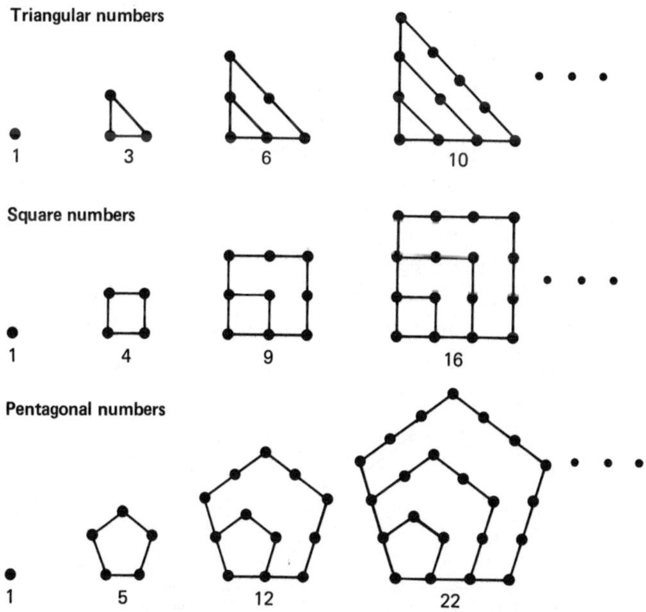

Fig. 1.10 Arranging numbers into geometric configurations

perhaps the first group of people to study what has since become known as number theory. In about 540 BC. they established a particularly pleasing and powerful way of looking at the nature of numbers. Many numbers can be arranged into geometric configurations (Figure 1.10).

It is interesting to develop formulae for these numbers. It is easy to see that the nth triangular number T_n looks like Figure 1.11.

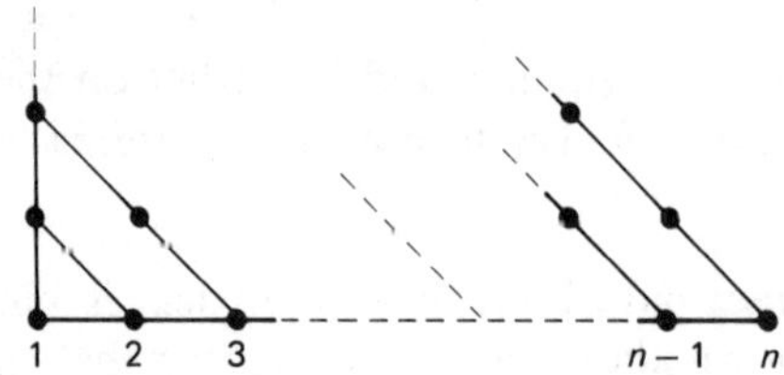

Fig. 1.11 Developing the pattern for the nth triangular number

In other words,

$$T_n = 1 + 2 + 3 + \ldots + n \qquad (1)$$

Using the same trick as we did in the magic squares we reverse this,

$$T_n = n + (n-1) + (n-2) + \ldots + 1 \qquad (2)$$

Adding (1) and (2) column by column, we obtain,

$$\begin{aligned} 2T_n &= (n+1) + (n-1+2) + (n-2+3) + \ldots + (n+1) \\ &= (n+1) + (n+1) + (n+1) + \ldots + (n+1) \\ &= n(n+1) \\ T_n &= n(n+1)/2 \end{aligned}$$

Q 1.7 (i) Calculate the 6th triangular number.
(ii) Draw a diagram to illustrate it.

Clearly, the nth square number S_n is given by,

$$S_n = n^2$$

The nth pentagonal number P_n is a little more difficult to formulate. It will look like Figure 1.12.

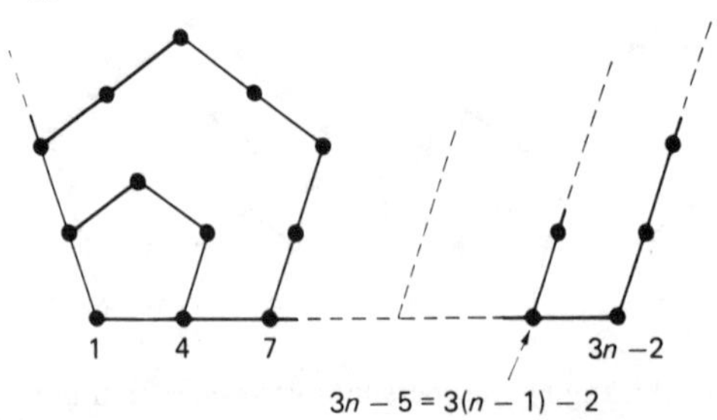

Fig. 1.12 The nth pentagonal number

Hence,

$$P_n = 1 + 4 + 7 + \ldots + (3n - 2) \qquad (1)$$

Reversing this sum, we have;

$$P_n = (3n - 2) + (3(n - 1) - 2) + (3(n - 2) - 2) + \ldots + 1 \qquad (2)$$

Adding (1) and (2) column by column gives

$$\begin{aligned} 2P_n &= (1 + 3n - 2) + (4 + 3(n - 1) - 2) + \ldots + (3n - 2 + 1) \\ &= (3n - 1) + (3n - 1) + (3n - 1) + \ldots + (3n - 1) \\ &= n(3n - 1) \\ P_n &= n(3n - 1)/2 \end{aligned}$$

Q 1.8 One might suppose that this formula is wrong because one has to divide by 2—how can we be sure that it divides exactly as required? To show that $n(3n - 1)$ does indeed divide exactly by 2 proceed as follows:

Suppose that n is *odd*. Then is $3n$ odd or even? So $3n - 1$ is? Hence, $n(3n - 1)$ is? This shows that $n(3n - 1)$ is divisible by 2.

Alternatively, suppose that n is *even*. Complete a similar argument.

Using these formulae and diagrams, some surprising results can very easily be established (Figure 1.13). This suggests that a square number is the sum of two consecutive triangular numbers. We can prove this very easily:

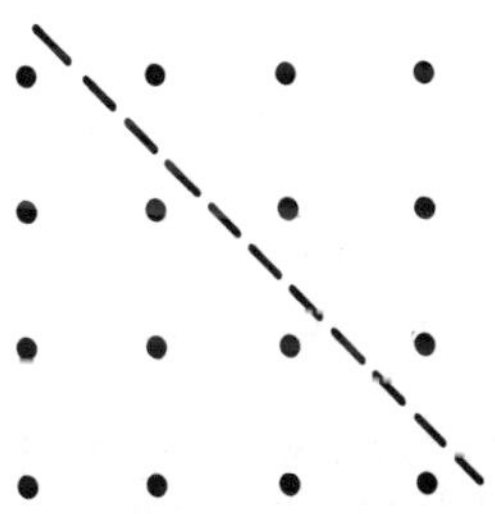

Fig. 1.13 A square number is the sum of two triangular numbers

$S_n = n^2$ is the nth square number. T_n, T_{n-1} are two consecutive triangular numbers, and

$$\begin{aligned} T_n + T_{n-1} &= n(n + 1)/2 + (n - 1)(n - 1 + 1)/2 \\ &= (n^2 + n)/2 + (n - 1)n/2 \\ &= (n^2 + n + n^2 - n)/2 \\ &= 2n^2/2 = n^2 \end{aligned}$$

Hence, $\quad T_n + T_{n-1} = S_n$

Q 1.9 A *rectangular* number is the number of dots contained within a rectangle. Can a rectangular number be prime?

Consider the diagram in Figure 1.14.

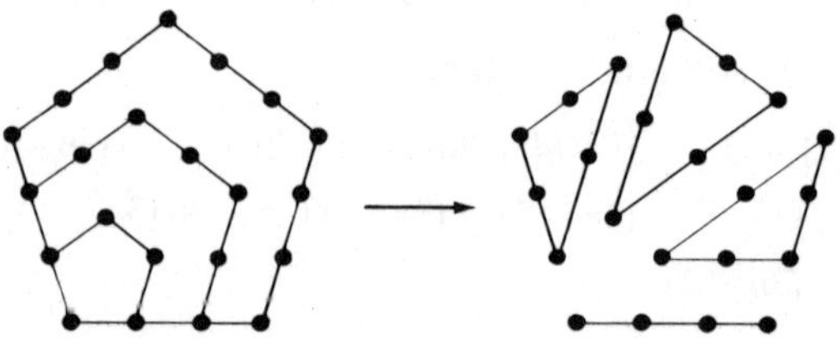

Fig. 1.14 The nth pentagonal number is the sum of three times the $(n-1)$th triangular number and n

This suggests that the nth pentagonal number is the sum of three times the $(n-1)$th triangular number and n. We can prove this as follows, $P_n = n(3n-1)/2$ is the nth pentagonal number.

$$\begin{aligned} 3T_{n-1} + n &= 3(n-1)(n-1+1)/2 + n \\ &= 3(n-1)(n)/2 + n \\ &= (3n^2 - 3n + 2n)/2 \\ &= (3n^2 - n)/2 \\ &= n(3n-1)/2 \end{aligned}$$

Hence, as required,

$$P_n = 3T_{n-1} + n$$

Q 1.10 Show geometrically and algebraically, that eight times any triangular number, plus one, is a square number.

It is of interest to generalize these sorts of numbers in two different directions. Let us first consider the hexagonal, septagonal, octagonal, ..., numbers. We then have the following formulae:

$$\begin{aligned} T_n &= n(n+1)/2 &&= (n^2+n)/2 \\ S_n &= n^2 &&= 2n^2/2 \\ P_n &= n(3n-1)/2 &&= (3n^2-n)/2 \\ H_n &= ? &&= ? \end{aligned}$$

Clearly, we shall very soon run out of letters to name these numbers. Let us denote the triangular numbers by 3_n, the square numbers by 4_n, etc. We then have,

$$3_n = (n^2 + n)/2$$
$$4_n = 2n^2/2$$
$$5_n = (3n^2 - n)/2$$

If we continued this pattern, can you see that we should have

$$6_n = (4n^2 - 2n)/2$$
$$7_n = (5n^2 - 3n)/2$$
$$\ldots$$

and in general, $M_n = \dfrac{(M-2)n^2 - (M-4)n}{2}$ the figure with M sides,

$$= n(Mn - 2n + 4 - M)/2$$

i.e. $$M_n = n(M(n-1) - 2(n-2))/2$$

Q 1.11 (i) How many dots does the third decagonal number have?
(ii) Draw a diagram to illustrate.

Another way of generalizing these numbers is to build three-dimensional shapes with spheres instead of dots (Figure 1.15).

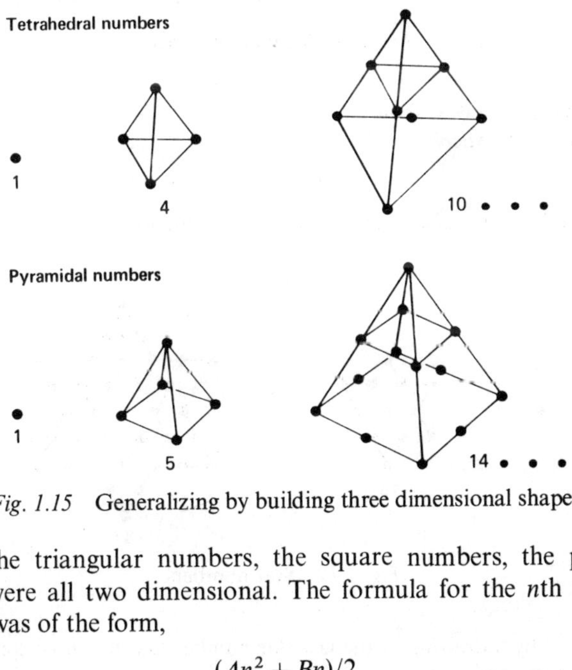

Fig. 1.15 Generalizing by building three dimensional shapes

Q 1.12 The triangular numbers, the square numbers, the pentagonal numbers were all two dimensional. The formula for the nth number of each type was of the form,

$$(An^2 + Bn)/2$$

(i) What form of formula would you expect for the three dimensional numbers—the tetrahedral numbers, the pyramidal numbers, ...?
(ii) In fact the formula is of the form,

$$(An^3 + Bn^2 + C)/6$$

Find the values of A, B, and C for the two cases we have considered.
(iii) See if you can generalize this to other three dimensional numbers, and generalize the formula.

Further investigations

1 Figure 1.16 shows a square number.

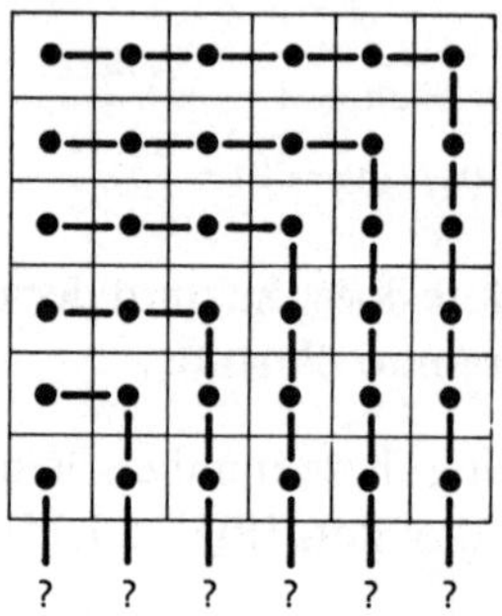

Fig. 1.16 A square number

(i) What else does it show?
(ii) What is S, where,

$$S = 1 + 3 + \ldots + (2n - 1)?$$

2 We can define the *star* numbers as shown in Figure 1.17;

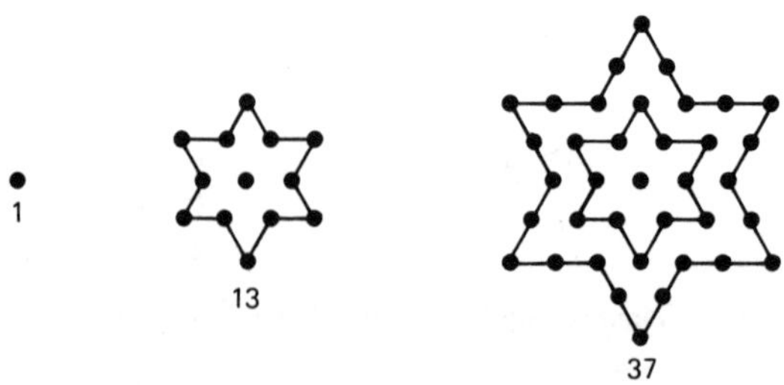

Fig. 1.17 Star numbers

(i) Show by a drawing of the last star number that it can be split up into 12 triangles and a dot.
(ii) Hence find a formula for the nth star number.

3 Consider the cube in Figure 1.18. It is desired to turn it into a *magic cube* by putting the numbers, 1, 2, 3, 4, 5, 6, 7, and 8, one to a side, in such a way that the total at each vertex is constant.

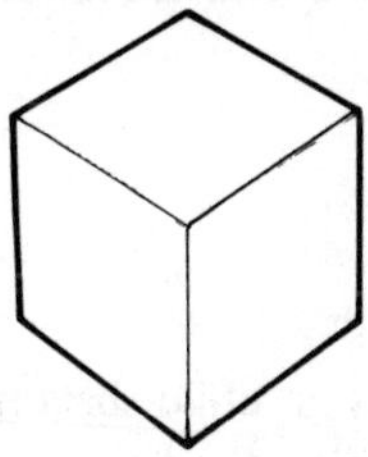

Fig. 1.18 A cube

Let the total at each vertex be t. Then the sum of all the vertices is $8t$. However, this total includes every number twice, since each side is connected to two vertices. Hence,

$$8t = 2(1 + 2 + 3 + \ldots + 8)$$

(i) What is the total at each vertex?
(ii) Hence show that it is impossible to construct a magic cube.

4 Investigate Question 3 for a tetrahedron.

5 Find a triangular number which is also square.

6 Using the formula you found in Question 2, $S_n = 6n(n-1) + 1$, show that three times a star number added to 2 always gives a number which is the sum of two consecutive squares.

$$\begin{aligned} \text{e.g.}\quad \text{When } n = 3, S_n &= 6 \times 3(3-1) + 1 \\ &= 37 \\ 3 \times 37 + 2 &= 113 \\ &= 8^2 + 7^2 \end{aligned}$$

7 Show that three times a square number added to 2 always gives a number which is the sum of three consecutive squares.

e.g. A square number is 16.

$$\begin{aligned} 3 \times 16 + 2 &= 50 \\ &= 3^2 + 4^2 + 5^2 \end{aligned}$$

8 Find a number which is the sum of two consecutive squares *and* three consecutive squares.

2 Patterns in mathematics

2.1 Tessellations

One of the first things that a child becomes aware of in the world surrounding him is pattern. Colour and shape are immediately recognizable, and in certain configurations, very pleasing. It is difficult to say exactly what pleases about such patterns and shape, but certainly, the order and regularity satisfy a very fundamental need.

Perhaps the simplest, and most satisfying shape is the circle.

> 'No wonder they call it a circle—it's so round. Notice how the inside comes precisely to the line and not one whit further. And how the outside can't possibly get in. No corners is one of the principal things about a circle. An oval has no corners, too, but they are not nearly as no corners as a circle has. Circles are nice because we can go round in them. Hardly anybody ever goes round in squares. Every single place on the outside of a circle is the same distance from the centre as every other place. You can't say that about a parallelepiped.
>
> Circles are often used to designate no runs on baseball score-boards, and enough of them after some figures will show the national debt in round numbers.
>
> There is nothing at the present time as round as a circle. But what with science is accomplishing nowadays, doubtless there'll be something rounder before long. I think circles would be a lot less dull if they were oval.'

Colonel Stoopnagel, writing in *The Saturday Evening Post*

One way of trying to come to terms with the pleasure that such a shape affords us is to talk of its *symmetry*. It is obvious that a circle has more symmetry than any other plane shape, and perhaps this is what Colonel Stoopnagel was really trying to say.

Using the circle as a starting point we can construct other shapes which are equally pleasing to the eye. We form the *regular polygons*, by dividing the circumference, or the angle at the centre, into 3, 4, 5, ... equal parts (Figure 2.1).

There is a mathematical pattern here as well as an artistic one,

3 sides $\longrightarrow$ each angle is 60°
4 sides $\longrightarrow$ each angle is 90°
5 sides $\longrightarrow$ each angle is 108°
6 sides $\longrightarrow$ each angle is 120°

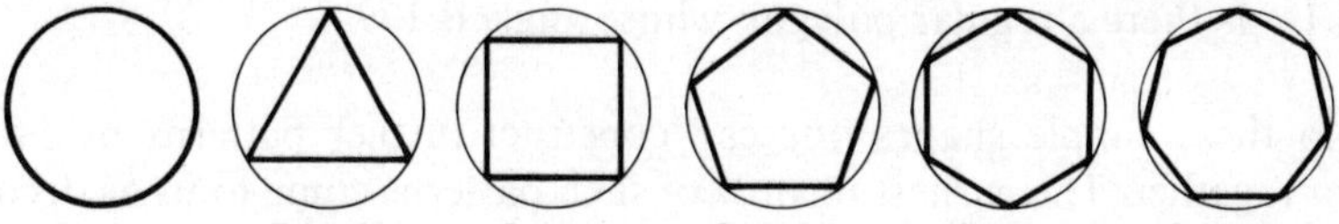

Fig. 2.1 Construction of polygons from a circle

If we constructed the polygon with n sides, what would be the size of each angle? In other words,

$$n \text{ sides} \longrightarrow \text{each angle is ?}$$

Actually, the pattern is,

$$n \longrightarrow \frac{(n-2)180^\circ}{n}$$

We can prove this in the following way. Consider a polygon with n sides. Label the vertices of the polygon A, B, C, D, … and the centre of the circle as O (Figure 2.2).

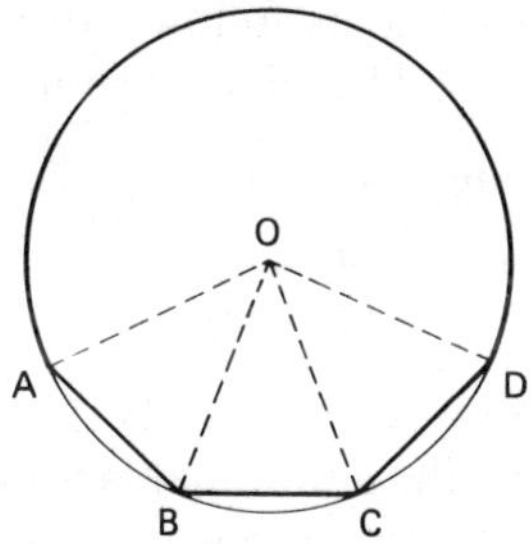

Fig. 2.2 Each side of the polygon is part of an isosceles triangle

Can you see that each of the angles, AOB, BOC, COD, … is $360/n^\circ$? But each of the triangles AOB, BOC, COD, … is isosceles, since the radii OA, OB, OC, … are all equal.
Hence,

$$\angle OAB = \angle OBA = \frac{180^\circ - 360/n^\circ}{2}$$

$$= \frac{180n - 360}{2n} = \frac{90n - 180}{n}$$

Now each angle of the polygon is composed of two of these angles,

$$\angle ABC = \angle OBA + \angle OBC$$

$$= \frac{90n-180}{n} + \frac{90n-180}{n} = \frac{180n-360}{n}$$

$$= \frac{(n-2)180^\circ}{n}$$

Q 2.1 Is there a regular polygon whose angle is 179°?

From these simple shapes one can construct further patterns by fitting them together. The earliest records of such patterns come to us as decorations—on floors, walls, ceilings, and other objects. Usually they occur in the form of mosaics or *tessellations*. A mosaic is simply a way of filling a plane area by the use of polygons. The three simplest are illustrated in Figure 2.3.

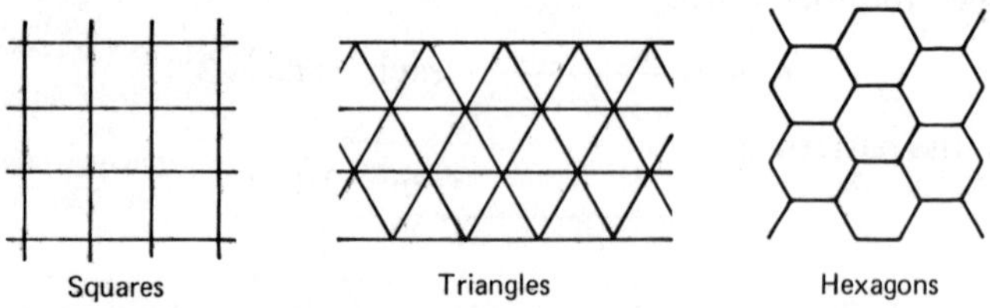

Fig. 2.3 Square, triangular, and hexagonal mosaics

Are these the only polygons that tessellate? We can in fact show that no other polygon will tessellate as follows. Consider the point at which several of these polygons meet. Then if we have m polygons at this point,

$$\frac{m(n-2)180^\circ}{n} = 360^\circ$$

$$\Rightarrow \quad m(n-2) = 2n \qquad (1)$$

This equation on the face of it is insoluble, because we have two unknowns, n and m. However, we have the additional fact that both n and m must be natural numbers. (Why?) We re-arrange (1) as follows and then try some 'guesstimates'.

$$m(n-2) = 2n \Rightarrow mn - 2m - 2n = 0$$

$$\Rightarrow \quad mn - 2m - 2n + 4 = 4 \text{ (adding 4 to both sides)}$$

$$\Rightarrow \quad m(n-2) - 2(n-2) = 4$$

$$\Rightarrow \quad (m-2)(n-2) = 4 \qquad (2)$$

Let us try the guess of $m = 1$:

$$(1-2)(n-2) = 4 \Rightarrow n = -2$$

This suggests that we can tessellate the plane with one polygon of -2 sides! Clearly it is not a possible solution.
Let us try $m = 2$:

$$(2-2)(n-2) = 4 \Rightarrow n = ?$$

This equation isn't even soluble!
Let us try $m = 3$:

$$(3-2)(n-2) = 4 \Rightarrow n = 6$$

Can you see that this is our tessellation of hexagons? That is, three polygons with six sides meeting at a point.
Again, we might try, $m = 4$:

$$(4-2)(n-2) = 4 \Rightarrow n = 4$$

This is our tessellation of squares.
If $m = 5$:

$$(5-2)(n-2) = 4 \Rightarrow n = 3\tfrac{1}{2}$$

Obviously, we cannot have a polygon with $3\frac{1}{2}$ sides, so this is not a possible solution.
If $m = 6$:

$$(6-2)(n-2) = 4 \Rightarrow n = 3$$

Q 2.2 What tessellation does this represent?

Try $m = 7$:

$$(7-2)(n-2) = 4 \Rightarrow n = 2\tfrac{4}{5}$$

Again this is not possible.
Can you see that if $m > 6$, then n will always be fractional?

Q 2.3 Supply a proof of this last statement.

Hence the only polygons that tessellate are the three already illustrated.

This proof is based upon the assumption that each of our polygons meet vertex to vertex. Can we form tessellations where, perhaps, a vertex of one polygon meets a side of another? Figure 2.4 shows that this is at least possible for squares. Is it possible for any other polygons?

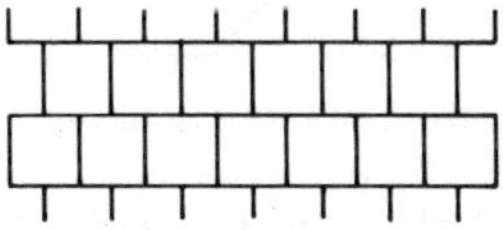

Fig. 2.4 A square tessellation where a vertex meets a side

We can use a similar analysis to that already adopted, to answer this question. Let m polygons meet along one side of a polygon; then if the polygons have n sides, we have

$$\frac{m(n-2)180^\circ}{n} = 180^\circ \text{ (angles on a straight line)}$$

$$\Rightarrow m(n-2) = n \Rightarrow mn - 2m = n$$
$$\Rightarrow \quad mn - 2m - n = 0$$
$$\Rightarrow \quad mn - 2m - n + 2 = 2$$
$$\Rightarrow \quad m(n-2) - (n-2) = 2$$
$$\Rightarrow \quad (m-1)(n-2) = 2 \qquad (3)$$

It is not worth bothering with $m = 1$ (why ?) so we begin, $m = 2$:

$$(2-1)(n-2) = 2 \Rightarrow n = 4$$

This is our displaced square tessellation.
When, $m = 3$:

$$(3-1)(n-2) = 2 \Rightarrow n = 3$$

Q 2.4 Draw the tessellation corresponding to this result.

Try $m = 4$:

$$(4-1)(n-2) = 2 \Rightarrow n = 2\tfrac{2}{3}$$

This is obviously not possible.

Q 2.5 Are there any other tessellations of this type?

We can construct other tessellations if we now admit the possibility of having more than one type of polygon. Perhaps the simplest of these is illustrated in Figure 2.5.

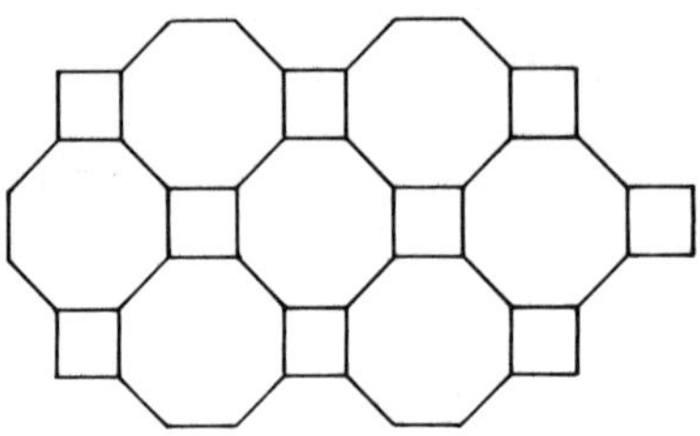

Fig. 2.5 Tessellation of more than one type of polygon

This is a tessellation of squares and octagons. Let us try and find all the tessellations of two polygons. As before we shall assume that they meet vertex to vertex. Suppose that we have m polygons with n sides and k polygons with h sides, all meeting at a vertex. Then, we have;

$$\frac{m(n-2)180^\circ}{n} + \frac{k(h-2)180^\circ}{h} = 360^\circ$$
$$\Rightarrow \quad \frac{m(n-2)}{n} + \frac{k(h-2)}{h} = 2$$
$$\Rightarrow \quad mh(n-2) + kn(h-2) = 2nh \qquad (4)$$

The solution of this equation is extremely difficult. However, we can use a similar technique to that already adopted. Let us assume first of all that we have one of each type of polygon meeting at a vertex, i.e. $m = 1$ and $k = 1$. Then,

$$\begin{aligned} & h(n-2) + n(h-2) = 2nh \\ \Rightarrow \quad & hn - 2h + nh - 2n = 2nh \\ \Rightarrow \quad & -2h - 2n = 0 \\ \Rightarrow \quad & n + h = 0 \end{aligned}$$

Can you see that this is impossible, as we should have a polygon with a negative number of sides?

Let us try two polygons of one type and one of the other. We have two possible cases to consider:

(a) $m = 1$ and $k = 2$

$$h(n-2) + 2n(h-2) = 2nh$$
$$hn - 2h + 2nh - 4n = 2nh$$
$$hn - 2h - 4n = 0$$

(b) $m = 2$ and $k = 1$

$$2h(n-2) + n(h-2) = 2nh$$
$$2hn - 4h + nh - 2h = 2nh$$
$$nh - 2n - 4h = 0$$

Can you see that these equations are basically the same? Hence, we need only consider one of them. Let us try some numbers in turn. Since n and h stand for the number of sides of polygons, we need only consider values greater than 2.

Let $n = 3$:

$$3h - 2h - 12 = 0$$
$$\Rightarrow \quad h = 12$$

This is equivalent to the solution, $n = 12$ and $h = 3$.

This tessellation is illustrated, in Figure 2.6.

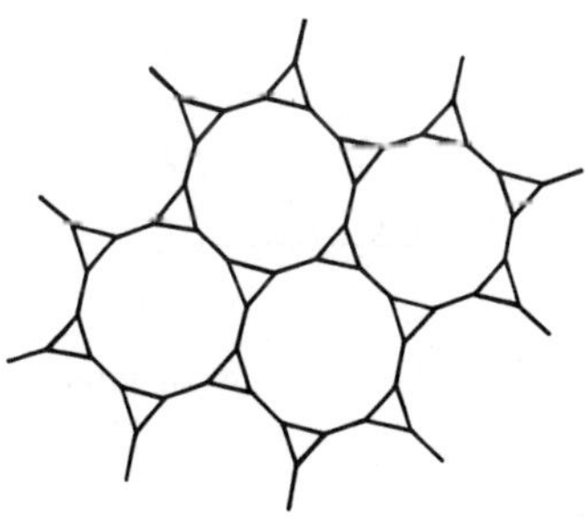

Fig. 2.6 Tessellation for $n = 12$ and $h = 3$

If we let $n = 4$:

$$4h - 2h - 16 = 0$$
$$2h = 16$$
$$\Rightarrow \quad h = 8$$

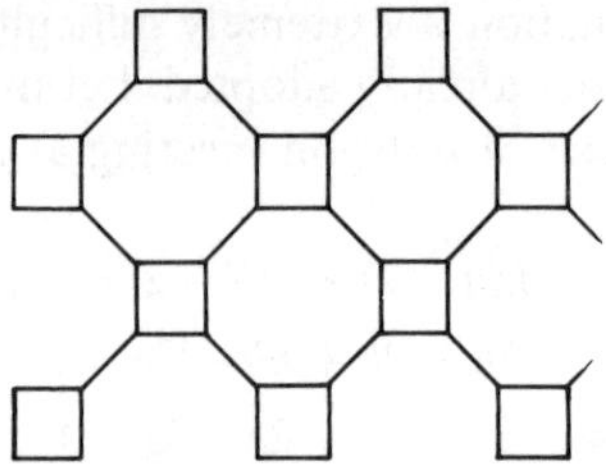

Fig. 2.7 Tessellation for $n = 4$ and $h = 8$

This yields the tessellation in Figure 2.7, with which we started. It is obvious that we are not going to get far in solving equation (4) in this way. In fact it might appear that there are a large number of solutions. However, there are only eight solutions. These are illustrated in Figure 2.8, together with the corresponding values of n, m, k, and h:

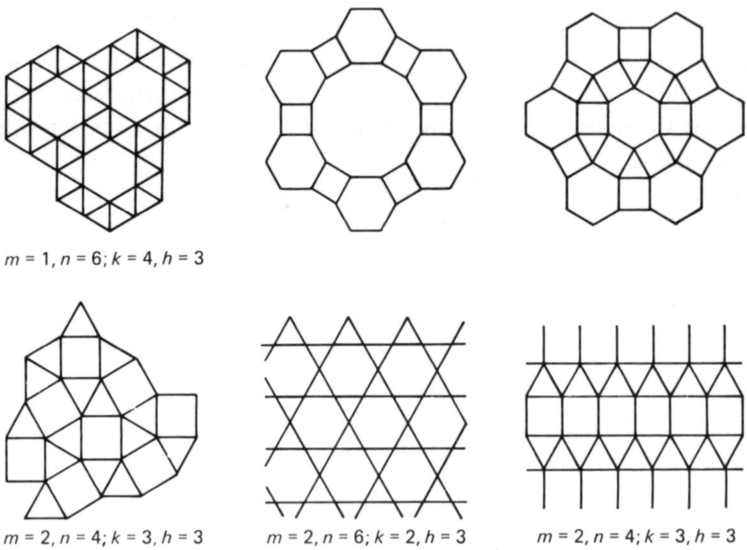

Fig. 2.8 Other solutions of equation 4

Some of the tessellations in Figure 2.8 have been left unmarked. This is because they consist of more than two types of polygons. It is a remarkable fact that no tessellation is possible with more than three polygons.

Although the number of tessellations of regular polygons is limited, no such limitation exists if we consider non-regular polygons. It is also possible to adopt some of the regular ones so that a more artistic effect is achieved. Figure 2.9 gives some examples of this.

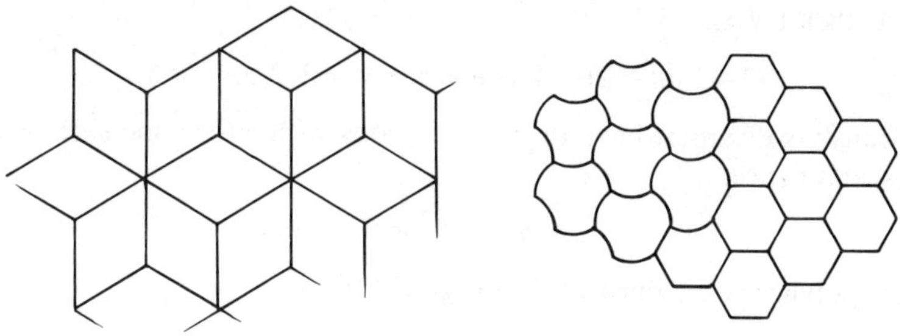

Fig. 2.9 Adaptation of tessellations with regular polygons

Interesting colour schemes can be designed for all the tessellations, and it is often just as rewarding colouring them as drawing them. Colour also has the effect of enhancing the essential geometric properties of the pattern, and occasionally, different colour schemes give rise to different features of the same pattern.

2.2 Polyhedra

An obvious generalization of the tessellations of the plane is the construction of solids, each of whose faces is a regular polygon. Such a solid is called a *polyhedron*. The various types of polyhedra have exercised a great fascination over the minds of mathematicians of all ages, among them some of the greatest names in mathematics.

The simplest polyhedra one can imagine are those made up of only one regular polygon. In fact the work we have already done on tessellations applies with only one modification to the work we need to do here. Consider one vertex of such a polyhedron, where we shall assume m regular polygons of n sides meet, vertex to vertex. Clearly, the sum of the angles of each of these polygons must not exceed 360°, otherwise they could not form a 'solid angle'. Hence,

$$\frac{m(n-2)180^\circ}{n} < 360^\circ$$

Can you see that this is precisely the same equation as (1) in the sections on tessellations, with the $=$ replaced by $<$?

Hence, we may write down immediately,

$$(m-2)(n-2) < 4 \tag{5}$$

In much the same way as before we shall try to solve this inequality in natural numbers n and m.

Q 2.6 Why need we not bother with $m = 1$ or $n = 2$?

Let us then try, $m = 3$:

$$(3-2)(n-2) < 4 \Rightarrow n < 6 \Rightarrow n = 1, 2, 3, 4, \text{ or } 5.$$

Obviously we cannot have regular polygons with either one side or two sides, and hence,

$$n = 3, 4, \text{ or } 5$$

These polyhedra are illustrated in Figure 2.10.

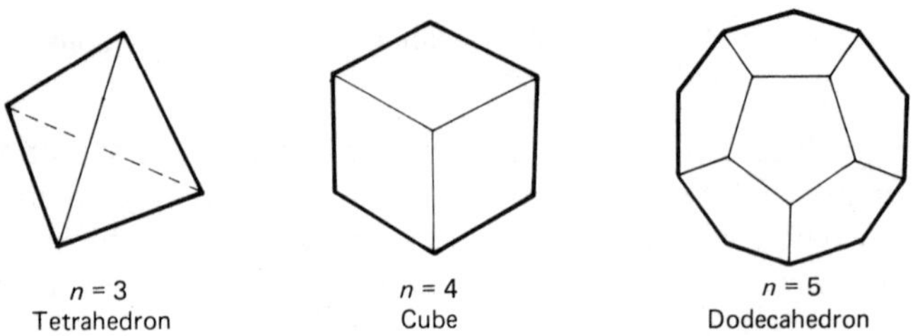

Fig. 2.10 Polyhedra with $n = 3$, $n = 4$, and $n = 5$

If $m = 4$:

$$(4-2)(n-2) < 4 \Rightarrow n < 4 \Rightarrow n = 1, 2, \text{ or } 3$$

Hence, $n = 3$.

Q 2.7 What polyhedron does this represent?

If $m = 5$:

$$(5-2)(n-2) < 4 \Rightarrow n < 3\tfrac{1}{3} \Rightarrow n = 1, 2, \text{ or } 3.$$

Obviously, $n = 3$.

This polyhedron is illustrated in Figure 2.11,

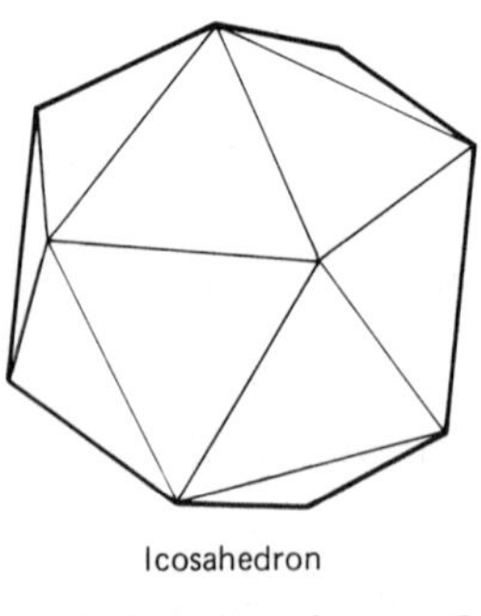

Fig. 2.11 Polyhedron for $m = 5$, $n = 3$

If $m = 6$:

$$\Rightarrow \quad (6-2)(n-2) < 4$$
$$\Rightarrow \quad n < 3$$
$$n = 1, \text{ or } 2$$

This is clearly impossible.

Q 2.8 Prove that these are the only regular polyhedra.

This group of polyhedra was known to Euclid in 300 BC. and the fact that they number five only, assured them of a place in mysticism. The Greeks associated them with the four supposed elements and the Universe. Of the five solids, the tetrahedron has the smallest volume for its surface, the icosahedron the largest; these therefore represent fire and water. The cube standing firm on its base corresponds to the stable earth. The octrahedron, which revolves freely when held by opposite vertices, corresponds to the mobile air. Finally, the dodecahedron with 12 sides, corresponds to the universe, because the zodiac has twelve signs. These five polyhedra are called the *Platonic solids*.

To construct models of these polyhedra is a fascinating and rewarding pastime. They are best made out of stiff paper or thin card, and a fast-drying glue is essential. The plans that follow (in Figures 2.12 and 2.13) for the models are called their *nets*. What are not shown are the tabs for securing one face to another. However, since each of these nets contains each side of a model twice, tabs need only be added to the even edges numbered from some convenient starting point. If small models are contemplated, it is likely that a supply of paper clips will come in handy, to hold together, while glueing, faces which the fingers can't.

The quality of the finished model is decided by only two things: the accuracy of its construction and the way it is painted. As regards accuracy, time spent on the actual drawing of the net is more than amply rewarded

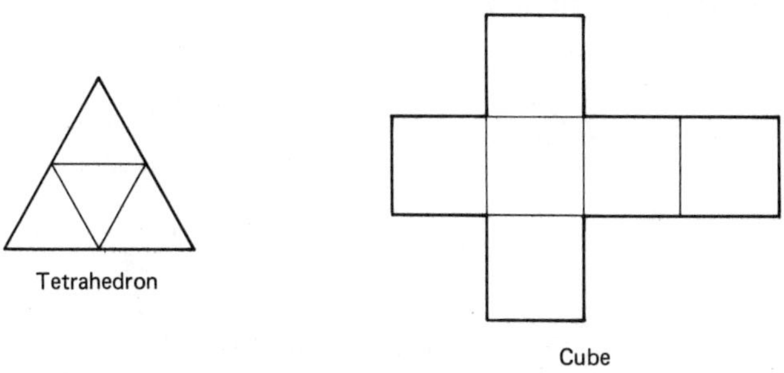

Fig. 2.12 Nets (plans) for construction of model tetrahedron and cube

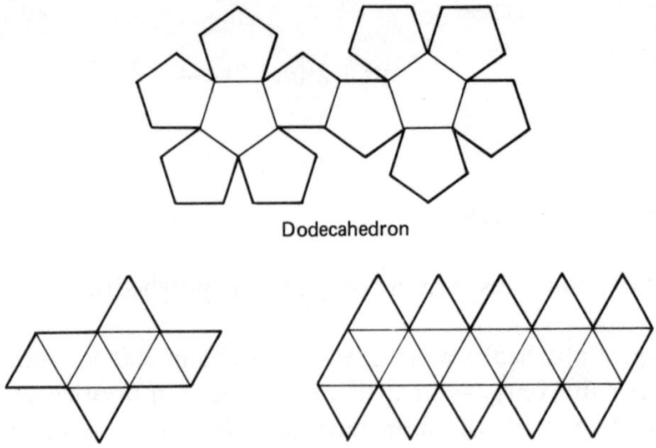

Fig. 2.13 Nets for more complicated polyhedra

in the satisfaction of getting the last face to fit perfectly. Colour schemes intelligently applied can greatly enhance the attraction of the finished models. They can also demonstrate some of the underlying geometry of the polyhedra.

Q 2.9 Why is the cube sometimes called a hexahedron?

Q 2.10 Let E be the number of edges, F the number of faces, and V the number of vertices of a polyhedra. For each of the Platonic solids, evaluate $F + V - E$.

Just as tessellations of the plane can be constructed with more than one regular polygon, so polyhedra can be similarly constructed. They are called the *Archimedean solids*. However, as distinct from the tessellations, there are an infinite number of them. In order to find them we shall have to solve inequalities similar to equation (4) of the last section. Before we attempt this, however, we shall eliminate two obvious groups, and concentrate on the remaining finite number.

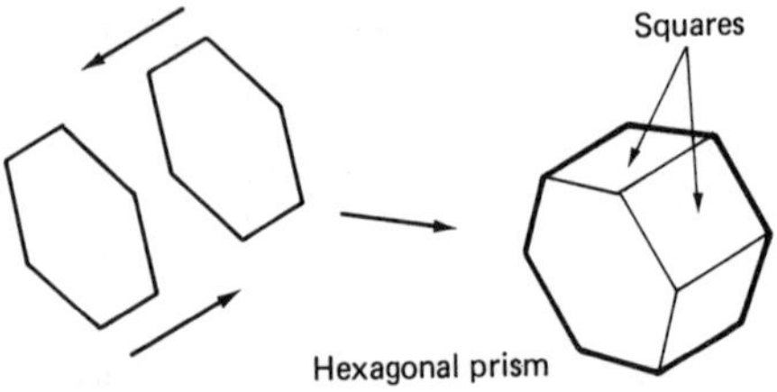

Fig. 2.14 Construction of hexagonal prism

One of these obvious groups is the *prisms*. A prism consists of two identical regular polygons similarly placed in parallel planes, with corresponding vertices joined by edges, so that the joining edges form sides of a square. This is illustrated in Figure 2.14 for two hexagons. Clearly, there are as many of these as there are regular polygons—i.e. an infinite number. However, they are not particularly pleasing.

Another group is the so called *antiprisms*. In these there are also two identical regular polygons placed in parallel planes, but this time, one polygon is twisted so that each of its vertices is midway between two vertices of the other, to each of which it is joined. These side faces are thus, equilateral triangles. This is illustrated in Figure 2.15 again for the hexagon.

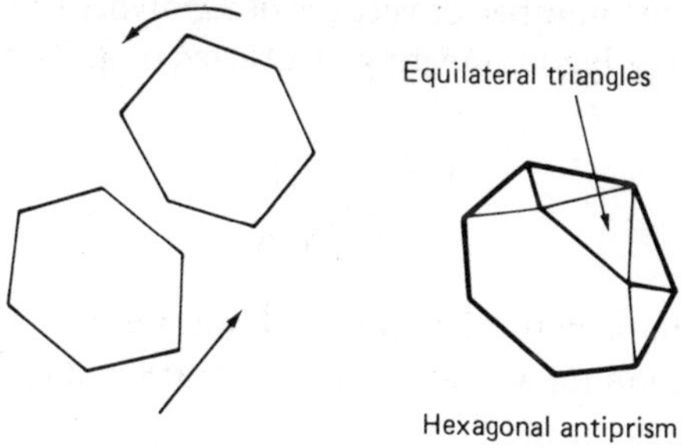

Fig. 2.15 Construction of hexagonal antiprism

If we ignore these two groups of Archimedean solids, there are only 13 others. Let us first consider those solids with two different regular polygons as faces. Then we need to solve the inequality,

$$mh(n-2) + kn(h-2) < 2nh$$

Q 2.11 Why is it not worth while trying $m = 1$ and $k = 1$ in this inequality?

Let us therefore try two polygons of one type and one polygon of another type meeting at a vertex, i.e. $m = 1$ and $k = 2$.

$$h(n-2) + 2n(h-2) < 2nh$$

$$\Rightarrow \quad hn - 2h + 2nh - 4n < 2nh$$

$$\Rightarrow \quad hn < 2(h + 2n)$$

Here, n and h represent the number of sides of each of these polygons. Let us begin with say, $n = 3$, i.e. a triangle as one of the polygons. Then,

$$3h < 2(h + 6)$$

$$\Rightarrow \quad h < 12$$

Hence, $n = 3 \Rightarrow h = 1, 2, 3, \ldots 11$

These polyhedra are composed of one triangle and two polygons of r sides, where r is in turn, $3, 4, 5, \ldots, 11$.

Unfortunately, not all of these solutions are solutions at all. The reason is not too difficult to find either. The problem can be stated as follows. When we considered the tessellations we proved that equation (4) was true for a tessellation. Conversely, it is easy to see that if equation (4) is true, then we have a tessellation. We say that equation (4) is both a *necessary* and a *sufficient* condition for the tessellation we are considering. However, the corresponding inequality for polyhedra is only necessary—it is not sufficient. What this means is this: solutions to our inequality might make polyhedra, but anything which is not a solution definitely won't.

There is an extra condition to be satisfied to ensure that we obtain polyhedra. If V is the number of vertices of a polyhedron and O is the sum of the angles of the polygons meeting at a vertex of it, then;

$$V = \frac{4}{2 - \dfrac{O}{180}} \tag{6}$$

Clearly, V must be a natural number. If we apply the solutions of our inequality to equation (6) we have both necessary and sufficient conditions for a polyhedron.

Q 2.12 Show that when $n = 3$ and $h = 5$, we do not get a polyhedron.

Q 2.13 Show that when $n = 3$ and $h = 4$, we obtain a prism. (*Hint*: use equation (6))

Q 2.14 What polyhedron do we obtain when $n = 3$ and $h = 3$?

It would be a long job to find all the solutions to the appropriate inequality and then check them in equation (6). Instead we shall give a sketch and the net of the 13 possible solutions. These are not arranged in their order of solution, but in the order of practical difficulty in their construction (the solids whose name includes the word *truncated* can all be obtained by cutting off the corners of some other solid).

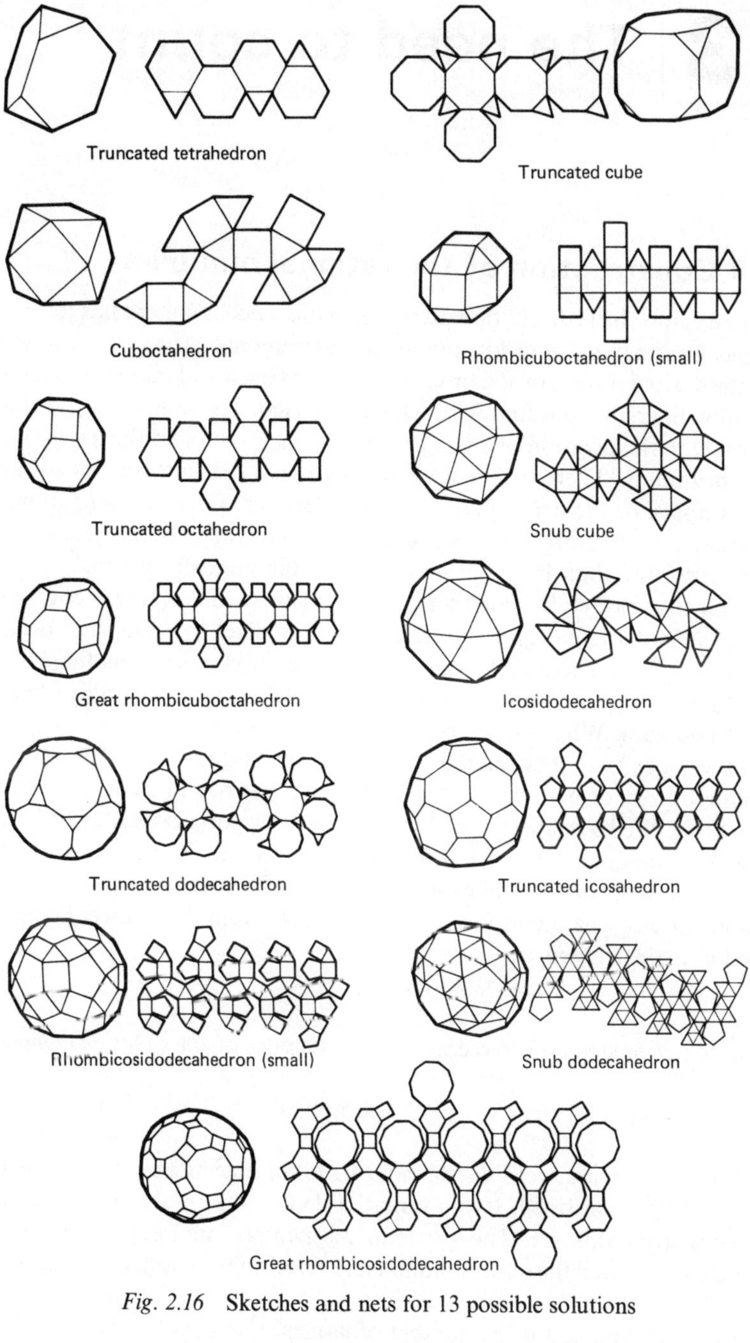

Fig. 2.16 Sketches and nets for 13 possible solutions

3 The need to count

3.1 Construction of the rational numbers

There can be little doubt that all mathematical development has its psychological roots in more or less practical requirements. However, once started in the desired direction, it gains momentum of itself and takes on a character totally different from that with which it started. The mathematical student is often in the position of a late comer to a race—he sees the finish but not the beginning. There is thus a tendency to view mathematics as a juggling of symbols, or the art of playing abstract games. These are part of mathematics, but they are one very small part. Nowhere is this more clearly demonstrated than in the development of the number system.

The inspiration for numbers is not hard to determine—counting is such a fundamental and physical operation that it must have occurred very early in the ascent of man. Just as the child counts his fingers, ears, and toes, so man counted his possessions. We can thus speak of five toes and two ears. What do we mean, though, when we speak of *two* fingers and *two* toes? Are they related? The vital step, and the crucial development in the history of man, is the realization that the *two* in both these cases is the same. We have abstracted the concept of *two* objects. Later still, comes another blinding discovery—that the idea of *two* is independent of the objects it describes. It makes sense to think of the *number two* as an object itself, and not as a quantifier of other objects. From these lowly beginnings we have constructed

$$2$$

It is no major step now to recognize the existence of, *the counting numbers*,

$$\{1, 2, 3, 4, 5, 6, \dots\}$$

For very obvious reasons we call this set of objects the *natural numbers*, and denote them by N. It is a remarkable fact that much of mathematics follows from this set. The German mathematician Leopold Kronecker said, 'God created the natural numbers, everything else is man's handiwork.'

Q 3.1 Can you count the number of natural numbers?

The everyday use of this set of numbers soon suggested two (there's that number again!) improvements—the operations of *addition* and *multiplication*. Clearly by putting together two (!) sets of the same objects, we obtain a new set of the same objects. The numbers attached to these sets as quantifiers must therefore be related, and, in turn, we may conceive of the numbers themselves as being related. Thus (Figure 3.1)

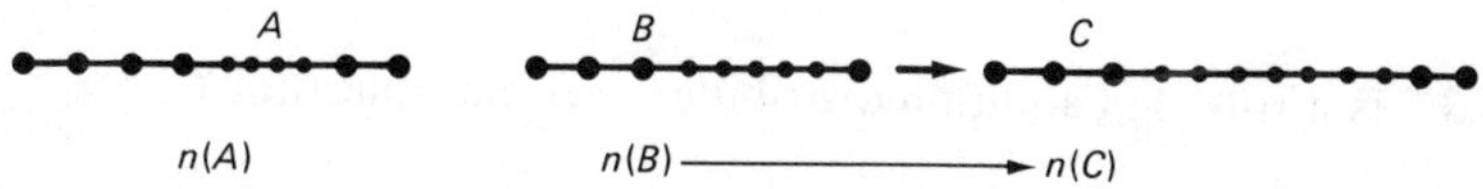

Fig. 3.1 Addition of natural numbers

and we can write,

$$a + b = c$$

Q 3.2 Try to explain what the symbol a above is:
(i) Is it a number?
(ii) Is it an abstraction of the concept of number?

Having established addition, we might try and investigate some of its properties. What do we mean by,

$$a + a + a + a + a + a?$$

Now we know how to count, and there are six symbols appearing in this expression. We can therefore agree to represent it by,

$$6a$$

and we now have another operation defined between natural numbers. It does not take us long to establish the following properties of the natural numbers with these two operations. For all a, b, and $c \in N$ we have:

(a) $a + (b + c) = (a + b) + c$ and $a.(b.c) = (a.b).c$
(b) $a.1 = 1.a = a$
(c) $a.(b + c) = a.b + a.c$
(d) $a + b = b + a$ and $a.b = b.a$

All these properties are so familiar that we often use them without a moment's thought. Indeed we often think of them as being obvious. It is easy to see, however, that they are not quite so obvious:

$$4.3 = 3 + 3 + 3 + 3$$

while $$3.4 = 4 + 4 + 4$$

Is it quite so obvious that they 'must' be the same? However, these rules are of fundamental importance in mathematics, because they keep on recurring. For this reason it is to our advantage to name them:

(*a*) The *associative* law.
(*b*) The *identity* element—there is a natural number which under multiplication leaves every other natural number unchanged.
(*c*) The *distributive* law—multiplication distributes over addition.
(*d*) Addition and multiplication are *commutative*.

Q 3.3 Is it true that addition distributes over multiplication?

From each of the operations + and . we can induce another operation. For instance, given the natural numbers 19 and 7, can we find a number which if added to 7 gives us 19? We clearly need to solve the equation

$$7 + x = 19$$

Intuitively, we recognize that we have to 'take away' 7 from 19. We say that the operation + induces the operation −, and write;

$$7 + x = 19 \longrightarrow x = 19 - 7$$
$$+ \longrightarrow -$$

In much the same way, we can consider the problem of 'sharing out' a number into a certain number of equal groups. Can we share 12 objects into four equal groups? If we can, we can conceive of the number 12 being shared by the number 4. Once more, intuitively, we can see that we seek the solution of the equation,

$$4 . x = 12$$

which leads us to the new operation

$$4 . x = 12 \longrightarrow x = 12/4$$
$$. \longrightarrow /$$

Let us deal with this new operation of *subtraction* first. The natural numbers are *closed* under the operations of addition and multiplication. That is, given any two numbers it is always possible to add and multiply them within the natural numbers. However, it is not always possible to subtract within the natural numbers, because some pairs of numbers take us outside this realm. For example,

$$4 - 1 = 3 \text{ (within the naturals)}$$

while

$$5 - 8 \text{ is not a natural number}$$

In order that we can perform this operation within the natural numbers we shall have to extend them. We shall have to invent 'new' numbers and adjoin them to the naturals, in much the same way that we invented the numbers in the first place. So long as the rules above still hold for the new system, we shall have solved our problem.

In order to perform subtraction, we are going to have to deal with problems like,

$$5 - 8$$

Let us assume for a moment that we have no idea what the number representing this might be. However, we certainly can represent it by the new symbol,

$$(5, 8)$$

We then note that,

$$(1, 4), (2, 5), (3, 6), (4, 7), \ldots, (n, n + 3), \ldots$$

must all be the same thing. This gives us a rule for deciding when two such numbers are the same, i.e. equal. Clearly we have;

$$(a, b) = (c, d)$$

whenever

$$a - b = c - d$$

We cannot state it in this form, however, since we don't yet know what subtraction is in the natural numbers. However, in view of the way that we induced subtraction from addition, we can interpret this as a corresponding addition. Thus,

$$(a, b) = (c, d)$$

whenever

$$a + d = c + b \qquad (1)$$

Next comes the problem of deciding how these 'numbers' shall be added and multiplied, bearing in mind that we want them to obey the same rules that we established for the natural numbers. Let us use what we already know of subtraction in the cases where it is defined in the naturals. Now,

$$7 - 2 = 5 \qquad \text{and} \qquad 8 - 4 = 4$$

hence, $$(7 - 2) \oplus (8 - 4) = 5 + 4 = 9$$

However, $$\begin{aligned}(7 - 2) \oplus (8 - 4) &= (7 + 8) - (2 + 4)\\ &= 15 - 6\\ &= 9\end{aligned}$$

Can you see that this means

$$(7 - 2) \oplus (8 - 4) = (7 + 8) - (2 + 4)$$

and so, in our new notation, we can write:

$$(7,2) \oplus (8,4) = (7+8, 2+4)?$$

Using this as a guide, we make a definition about the addition of these new numbers:

$$(a,b) \oplus (c,d) = (a+c, b+d) \qquad (2)$$

Q 3.4 (i) Which natural number does the new number $(2,1)$ behave like?
(ii) Which natural number does the new number $(n+1,1)$ behave like?

This question should have led you to the conclusion that the new numbers,

$$\{(2,1), (3,1), (4,1), \ldots (n+1,1) \ldots\}$$

represent the natural numbers. They obey the familiar rules. For instance, in the natural numbers;

$$3+5=8 \qquad \text{and so we have to check,}$$

$$(4,1) \oplus (6,1) = (9,1)$$

Now $\quad (4,1) \oplus (6,1) = (4+6, 1+1) \quad \text{by} \quad (2)$

$$= (10,2)$$

However $\quad (10,2) = (9,1) \quad \text{since } 10+1 = 2+9$

and we can then use (1).

We have made a distinction between two operations, $+$ and $\oplus$. The former refers to addition between natural numbers, and the latter between our new numbers (a,b). The distinction between these operations is real, but it would be pedantic to continue to differentiate. Henceforth we shall talk of the single operation $+$, and mean addition in both senses.

We have not yet decided how we are going to multiply these new numbers. Adopting the same approach as for addition, we note that,

$$(8-5) \odot (9-6) = 3.3 = 9$$

However,

$$\begin{aligned}(8-5) \odot (9-6) &= (8.9 + 5.6) - (8.6 + 5.9)\\ &= (72+30) - (48+45)\\ &= 102 - 93\\ &- 9\end{aligned}$$

Can you see that this means,

$$(8-5) \odot (9-6) = (8.9 + 5.6) - (8.6 + 5.9)$$

and so, in our new notation we can write:

$$(8,5) \odot (9,6) = (8.9 + 5.6, 8.6 + 5.9)?$$

Using this as a guide, we make a definition about the multiplication of these new numbers:

$$(a,b) \odot (c,d) = (a.c + b.d, a.d + b.c) \tag{3}$$

Again, we shall henceforth drop the distinction between the two operations . and $\odot$.

We now come to the crucial question—what are these new numbers (a,b) that we have created? Remember, that we invented them to solve the difficulty about subtracting in the naturals. Hence, we would expect this new set of numbers to be *the integers*,

$$\{\ldots -3, -2, -1, 0, 1, 2, 3, \ldots\}$$

which we shall denote by J.

To check that is indeed the case, we investigate several of their properties.

Is there a zero? The characteristic of zero is that it is the identity for addition, i.e. for any number a,

$$a + 0 = 0 + a = a$$

It should be obvious that any new number of the form (k,k) is a zero because,

$$k - k = 0$$

Now, any new number can be written in the form (a,b) where a and b are natural numbers.

Hence, $(a,b) + (k,k) = (a+k, b+k)$ by (2)

However, $(a,b) = (a+k, b+k)$ because

$a + (b+k) = b + (a+k)$ and then (1)

Hence, $(a,b) + (k,k) = (a,b)$ and so (k,k) is the addition identity.

Furthermore, we have:

$$(a,b).(k,k) = (a.k + b.k, a.k + b.k) \text{ by (3)}$$
$$= (c,c)$$

Clearly, this is equivalent to,

$$a.0 = 0 \text{ in the integers}$$

Hence, any number of the form (k,k) behaves exactly as a zero in the integers. Consider the following,

$$(n+1, 1) + (1, n+1) = (n+2, n+2) \text{ by (2)}$$
$$= (k,k)$$

Can you see that this equivalent to,

$$n + (-n) = 0 \text{ in the integers?}$$

Moreover, we have:

$$(1,n+1).(1,n+1) = (1+(n+1)^2, n+1+n+1) \text{ by (3)}$$
$$= (n^2+2n+2,\ 2n+2)$$
$$= (n^2+1,\ 1) \text{ by (1)}$$

which is the same as,

$$(-n).(-n) = n^2 \text{ in the integers}$$

Thus we may safely conclude that these new numbers are the integers, because they behave in exactly the same way. We can now dispense with our clumsy, but necessary, symbolism and revert to the more traditional one. Thus, we shall write,

$$(2,3) \text{ as } -1 \quad \text{and} \quad (3,1) \text{ as } 2 \text{ etc.}$$

All this is summarized in Figure 3.2.

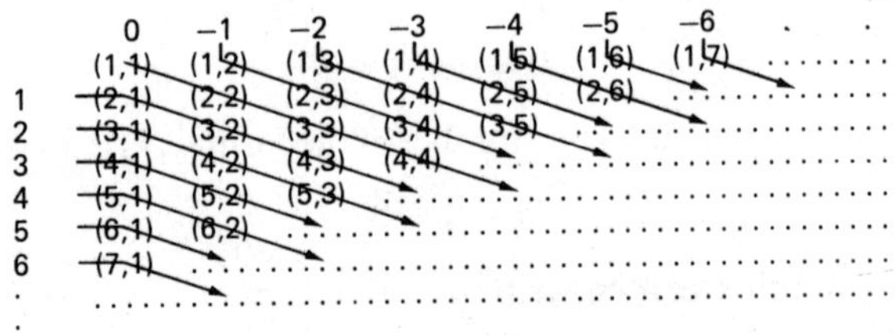

Fig. 3.2 More traditional symbolism

The only remaining thing to check is that we can really subtract any pair of these integers to give another integer. Consider the equation,

$$x+7=3 \text{ where 7 and 3 are natural numbers.}$$

We know that this equation is not satisfied by any natural number x because $7>3$. Hence, we shall have to broaden our horizons to find any possible solutions. We thus consider the integers J. In this set we have,

$$7 \longrightarrow (8,1),\ 3 \longrightarrow (4,1) \text{ and we can write,}$$
$$x \longrightarrow (a,b)$$

Our equation then becomes,

$$(a,b)+(8,1)=(4,1)$$
$$(a+8,b+1)=(4,1) \text{ by (2)}$$

and so, $$(a+8)+1=4+(b+1) \text{ by (1)}$$

Hence, $$a+9=b+5$$

and so, $$a+4=b$$

This means that the solution of our equation is $(a, a+4)$, which is obviously the integer -4, as expected.

Q 3.5 Is it always possible to solve the equation $x + a = b$, where a and b are integers, in the integers?

The extension of the natural numbers into the integers is achieved by requiring that the operations of addition and subtraction should be closed in this system. However, it does not guarantee that the operations of multiplication and division will be closed. For instance,

$$6/3 = 2$$

$$\text{and } 18/-9 = -2 \text{ are both integers}$$

However, $\quad 7/4 \quad$ is not an integer

In order to completely close the number system for the four basic operations, we shall have to extend the integers, just as we extended the natural numbers.

We are going to have to deal with problems like 7/4. Let us assume for a moment that we haven't a clue what this might be. We can with certainty, however, represent it by,

$$(7, 4)$$

By the very nature of 'sharing out' we then note that,

$$(7,4), (14,8), (21,16), \ldots, (7n, 4n), \ldots$$

must all be the same thing. This gives us a rule for deciding when two such numbers are the same, i.e. equal. Clearly we have,

$$(a, b) = (c, d)$$

whenever $\quad a/b = c/d$

We cannot state it in this form, however, since we don't yet know what division is in the integers. However, in view of the fact that we induced division from multiplication, we can interpret this as a corresponding multiplication. Thus,

$$(a, b) = (c, d)$$

whenever $\quad a.d = c.b \quad$ (1)

Next comes the problem of deciding how these 'numbers' shall be added and multiplied, bearing in mind that we want them to obey the same rules that we established for the natural numbers. Let us use what we already know of division in the cases where it is defined in the integers. Now,

$$12/6 = 2 \quad \text{and} \quad 15/5 = 3$$

Hence $\quad 12/6 \oplus 15/5 = 2 + 3 = 5$

However, $$12/6 \oplus 15/5 = \frac{5.12 + 6.15}{5.6} = \frac{60 + 90}{30}$$

$$= 5$$

Can you see that this means

$$12/6 \oplus 15/5 = \frac{5.12 + 6.15}{5.6}$$

and so in our new notation we can write:

$$(12, 6) \oplus (15, 5) = (5.12 + 6.15, 5.6)?$$

Using as a guide, we make a definition about addition of these numbers:

$$(a, b) \oplus (c, d) = (a.d + b.c, b.d) \tag{2}$$

Q 3.6 Which of these numbers corresponds to the integers? Which of them correspond to the natural numbers?

To find a corresponding rule for multiplication we note that,

$$12/6 = 2 \quad \text{and} \quad 15/5 = 3$$

Hence $$(12/6) \odot (15/5) = 2.3 = 6$$

However, $$(12/6) \odot (15/5) = (12.15/6.5) = (180/30) = 6.$$

Can you see that this means

$$(12/6) \odot (15/5) = (12.15/5.6)$$

and so in our new notation we can write:

$$(12, 6) \odot (15, 5) = (12.15, 5.6)?$$

Using this as a guide, we make a definition about multiplication of these new numbers:

$$(a, b) \odot (c, d) = (a.c, b.d) \tag{3}$$

As before, we shall drop the distinction between the extended operations $\odot$ and $\oplus$, and . and +.

We come now to the crucial question—what are these new numbers (a, b) where a and b are integers, that we have constructed? When we recall that we invented these numbers to solve the problem of division in the integers, it is easy to see that this set of new numbers must be *the rationals*, i.e. numbers of the form p/q where p and q are integers.

Q 3.7 Is it now always possible to solve the equation,

$$a.x = b$$

where a and b are integers, in the integers?

Is it possible to solve this equation in the rationals if a and b are rationals?

Actually, we haven't completely closed this system to division. A difficulty arises if we allow division by zero. To express this in the standard notation, it is always true that,

$$a.0 = b.0 \text{ for all rational numbers } a \text{ and } b.$$

If we allow division by zero, this would amount to,

$$a = b$$

Hence, we specifically exclude this particular case of division.

3.2 Field extensions

A very useful geometrical interpretation of the rational number system is constructed in the following way. On a straight line, we choose any initial point, and from this point we mark off a segment. We label the initial point 0, and the constructed point 1. (Figure 3.3).

Fig. 3.3 Construction of the unit length

This establishes the length of the segment from 0 to 1 as the *unit* length. The positive and negative integers are then represented as a set of equidistant points on this line, now called a *number axis*, the positive integers on the same side of 0 as 1, and the negative integers on the opposite side.

To represent fractions with a given denominator n, we divide each of the segments of unit length into n equal parts: the points of this division then represent the fractions with denominator n. If we do this for all integers n, then all the rational numbers will be represented by points of the number axis. We shall call such points, *rational points.*

There are now several interesting properties of the rational number system which we can establish. Firstly, between any two rational numbers, there is another. Our number axis makes this clear—the mid-point of any two points is clearly between them. All that we have to show is that this mid-point is itself rational. Suppose that we are given two rational numbers a/b and c/d. Then, obviously,

$$(a/b + c/d)/2$$

lies between them and is itself rational:

$$(a/b + c/d)/2$$

$$= \frac{ad + bc}{2bd} = r/s \text{ where } r \text{ and } s \text{ are integers.}$$

Another property is that given any segment of the number axis, no matter how small, we can always find a rational number inside it. Suppose that we were given the segment between, and including, the points A and B, which we denote by (A, B). Consider the segment $(0, 1/n)$. As n increases, the 'size' of this latter segment decreases, and it is intuitively obvious that by taking a large enough value of n, a segment $(0, 1/n)$ can be produced which is smaller than the segment (A, B). We may conclude that one of the fractions m/n must lie in the segment (A, B). It follows that there are as many rational numbers in a given segment as we care to construct. For, there is at least one rational point m/n in the segment (A, B) however, small. There are two more in the smaller segments, $(A, m/n)$ and $(m/n, B)$, and each of these is obviously in the segment (A, B). We can continue this construction endlessly, and the result follows.

This seems to suggest that the rational numbers completely cover the number line. This is not so, however, as the Pythagorean mathematicians discovered, much to their horror. Consider the number axis drawn in Figure 3.4, and the right-angled triangle constructed on it:

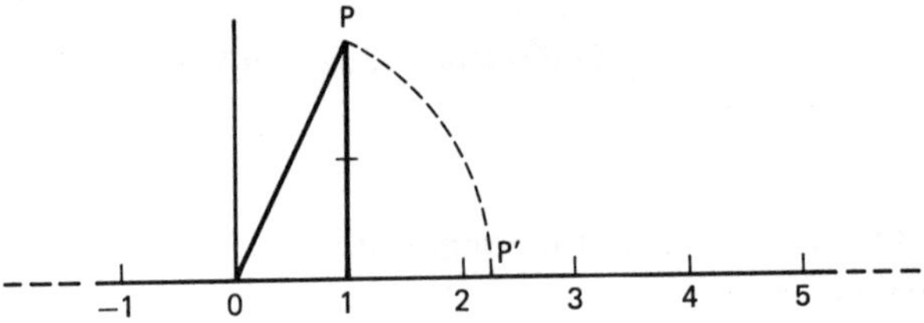

Fig. 3.4 Certain points on the number line are not rational

Triangle OP1 is such that (O, 1) is of unit length, and (P, 1) is twice the unit length. With centre O, we rotate the point P onto the number axis, giving the point P′. Using the theorem of Pythagoras, we have

$$\begin{aligned} \mathrm{OP}^2 &= 1^2 + 2^2 \\ \Rightarrow \quad \mathrm{OP} &= \sqrt{5} \\ \Rightarrow \quad \mathrm{OP}' &= \sqrt{5} \end{aligned}$$

Let us suppose that P′ is represented by a rational number, then

$$\begin{aligned} \mathrm{OP}' &= a/b \text{ for some integers } a \text{ and } b. \\ \Rightarrow \quad \sqrt{5} &= a/b \\ \Rightarrow \quad 5b^2 &= a^2 \end{aligned}$$

We may suppose that the fraction a/b has been reduced to its lowest terms (by the use of equation (1a)) and that a and b have then no common factors.

Now the left hand side (l.h.s.) of this equation is divisible by 5, and hence the r.h.s is, too. This means that a^2 is divisible by 5, and hence a is divisible by 5.

$$\Rightarrow \quad a = 5c \text{ for some integer } c$$

Hence,
$$5b^2 = a^2 \text{ now becomes, } 5b^2 = (5c)^2$$
$$\Rightarrow \quad 5b^2 = 25c^2 \Rightarrow b^2 = 5c^2$$

Now the r.h.s. of this last equation is divisible by 5, and so therefore, is the l.h.s. Hence b^2 is divisible by 5, and so b is divisible by 5.

We have now proved: if $\sqrt{5} = a/b$ then a is divisible by 5 and b is divisible by 5. However, a and b were reduced to their lowest terms, and so they cannot have a common factor of 5. We conclude that $\sqrt{5}$ cannot be expressed as a rational number. This means that there are points of the number line which are not rational, i.e. $\sqrt{5}, 2\sqrt{5}, 3\sqrt{5}, 4\sqrt{5}, \ldots$.

Q 3.8 Prove that $a\sqrt{5}$, where a is an integer, is not rational.

Q 3.9 Where does the proof that $\sqrt{4}$ is not rational break down?

The rational number system is called a *field* because it is closed under the four rational operations of addition, subtraction, multiplication, and division (except by zero). We shall denote the rational number field by R_0. Our development of the number system from the natural numbers N to the field R_0 proceeded at each stage because of the need to solve equations like,

$$x + a = b \quad \text{and} \quad ax = b$$

the first yielding the integers J, and the second the rational field R_0. However, there are still many equations which are not soluble in R_0, though all linear ones are. Consider the equation,

$$x^2 - 2 = 0$$

Clearly, $x = \sqrt{2}$ and it is easy to prove that $\sqrt{2}$ is not rational.

There is in fact, a very easy characterization of numbers, which separates them into rational and *irrational* categories. To see what this might be, let us consider some rational numbers. Any number of the form a/b, where a and b ($\neq 0$) are integers, is rational. By the ordinary process of long division it is easy to verify,

$$4/5 = 0.8 \quad \text{and} \quad 2/7 = 0.285\,714\,285\,714\,285\,714\ldots.$$

This, and many other examples, strongly suggest that all rational numbers can be represented either by a terminating, or a recurring decimal. We can prove this as follows. Consider the rational number a/b. If we change this number into its decimal by dividing b into a, either the process terminates or at each stage of the division there is a remainder to carry on to the next division. These remainders will be the integers between 1 and $(b-1)$, so that there are at most $b-1$ different possibilities for the value of the remainder. Hence, within at most b divisions, some remainder will be repeated, and thereafter, the process continually repeats itself in the same cycle.

We may just as easily prove the converse of this: all periodic decimals are rational numbers. (In this context, we shall regard a decimal that terminates as periodic—the zeros after the termination being the repetition.) The proof is extremely simple, but difficult to formulate in the general case; we shall content ourselves with a particular demonstration.

Consider the decimal, $x = 1.273\,414\,141\,414\ldots$

then $1000x = 1273.414\,141\,414\ldots$

and $100\,000x = 127\,341.414\,141\,41\ldots$

subtracting, $99\,000x = 126\,068.000\,000\,0\ldots$

i.e. $x = 126\,068/99\,000$ a rational number.

Thus we may say that a number is irrational if its decimal expansion is not periodic, after some finite number of other digits.

Because the rational field of numbers is insufficient for mathematics, in the sense that not all equations can be solved within it, one possibility is to extend it again, to include such solutions. Let us extend R_0 by adjoining the element $\sqrt{2}$ to it. If we call this new system $R_0(\sqrt{2})$, we can easily determine what properties it has. Consider the number line in Figure 3.5.

Fig. 3.5 Number line containing $\sqrt{2}$

From R_0 we can find all the rational points of this line, and from the single element $\sqrt{2}$, we can find all irrational points $a\sqrt{2}$ where a is also rational. Hence, all the elements of $R_0(\sqrt{2})$ are of the form

$$a + b\sqrt{2} \quad \text{where } a \text{ and } b \text{ are rationals}$$

The interesting property of $R_0(\sqrt{2})$ is that it too is a field; that is, it is closed under the four rational operations. We can prove this very easily. Suppose that a, b, c, and $d \in R_0$, then:

(a) *Addition*

$$(a + b\sqrt{2}) + (c + d\sqrt{2}) = (a + c) + (b + d)\sqrt{2}$$

$(a+c)$ and $(b+d)$ are rational, hence the sum of two elements of $R_0(\sqrt{2})$ is another element.

(b) *Subtraction* similarly, the difference of two elements of $R_0(\sqrt{2})$ is also an element.

(c) *Multiplication*

$$\begin{aligned}(a+b\sqrt{2})(c+d\sqrt{2}) &= ac+ad\sqrt{2}+bc\sqrt{2}+bd2\\ &= (ac+2bd)+(ad+bc)\sqrt{2}\end{aligned}$$

Now, $(ac+2bd)$ is a rational, and so is $(ad+bc)$. Hence, the product of two elements of $R_0(\sqrt{2})$ is another element.

(d) *Division*

$$\begin{aligned}\frac{(a+b\sqrt{2})}{c+d\sqrt{2}} &= \frac{(a+b\sqrt{2})(c-d\sqrt{2})}{(c+d\sqrt{2})(c-d\sqrt{2})}\\ &= \frac{(a+b\sqrt{2})(c-d\sqrt{2})}{c^2-2d^2}\end{aligned}$$

(N.B. We stated that we could not divide by zero in R_0, and so we must check that

$$c^2-2d^2 \neq 0$$

Now, if $$\begin{aligned}c^2-2d^2=0 &\Rightarrow c^2=2d^2\\ &\Rightarrow \sqrt{2}=c/d\end{aligned}$$

i.e. $\sqrt{2}$ is a rational number, which we have proved not to be true. Hence, the denominator, $c^2-2d^2 \neq 0$.)

Then $$\frac{(a+b\sqrt{2})}{(c+d\sqrt{2})} = \left(\frac{ac-2bd}{c^2-2d^2}\right)+\left(\frac{bc-ad}{c^2-2d^2}\right)\sqrt{2}$$

which is again an element of $R_0(\sqrt{2})$

Hence, $R_0(\sqrt{2})$ is a field. In fact this field is larger than the rational field R_0, which is a *subfield* of it. It is still not sufficient to cover completely the number line, since for example, $\sqrt{3}$ is not an element of it.

Q 3.10 Prove that $\sqrt{3} \notin R_0(\sqrt{2})$.

Q 3.11 The number 142 857 has the property that multiplication by any of the numbers 2, 3, 4, 5, or 6 gives a number which has the same digits as 142 857 but in a different order. By changing 1/7 into a decimal, try to explain this property.

Since the field $R_0(\sqrt{2})$ still has gaps, we need to extend it further. Let us do this by adjoining the new element $\sqrt{3}$. If we denote $R_0(\sqrt{2})$ by R_1, we shall then form $R_1(\sqrt{3})$. Any element of this extended field will be of the form

$$a + b\sqrt{3}$$

where a and b are elements of R_1, i.e.

$$a = p + q\sqrt{2} \quad \text{and} \quad b = r + s\sqrt{2} \text{ where } p, q, r, \text{ and } s \text{ are}$$

in R_0, the rationals.

It is again easy to verify that $R_1(\sqrt{3})$ is a field, which contains as a sub-field R_1. Let us denote the larger field $R_1(\sqrt{3})$ by R_2. Obviously, we can continue these sorts of extensions indefinitely, giving the sequence of fields,

$$R_0 \subset R_1 \subset R_2 \subset R_3 \subset R_4 \ldots R_{n-1} \subset R_n \subset \ldots$$

We call each of the fields R_n, a *square root extension field of the rationals.*

In this sequence of fields, it is tempting to suppose that eventually we shall reach a field R_n, say, which completely covers the number axis. Unfortunately this is not the case. Suppose that R_n is constructed from the field R_{n-1} by adjoining the element $\sqrt{c}$. Then, the general element of R_n is

$$a + b\sqrt{c}$$

where a, b, and $c \in R_{n-1}$, but $\sqrt{c} \notin R_{n-1}$ (otherwise there would be no need to adjoin it). If R_n completely covers the number axis, it contains every point, and in particular it must contain the point

$$\sqrt{(1 + \sqrt{c})}$$

If this point is in R_n, we then have,

$$\sqrt{(1 + \sqrt{c})} = p + q\sqrt{c} \quad \text{for } p \text{ and } q \in R_{n-1}$$

$$\Rightarrow \quad 1 + \sqrt{c} = p^2 + cq^2 + 2pq\sqrt{c} \tag{1}$$

$$\Rightarrow \quad \sqrt{c} = \frac{p^2 + cq^2 - 1}{1 - 2pq} \tag{2}$$

$$\Rightarrow \quad \sqrt{c} \in R_{n-1} \text{ since } p, q, \text{ and } c \text{ are.}$$

By our initial assumptions, this is not so, and hence (2) must be false. This can only be the case if we have divided by zero. (The non-divisibility of zero carries over with all the field extensions.)

$$\Rightarrow \quad 1 - 2pq = 0 \Rightarrow 2pq = 1 \tag{A}$$

Putting this result into (1) we obtain,

$$1 + \sqrt{c} = p^2 + cq^2 + 1\sqrt{c}$$

$$\Rightarrow \quad p^2 + cq^2 = 1 \tag{B}$$

Subtracting (A) from (B) we obtain,

$$p^2 - 2pq + cq^2 = 0$$
$$\Rightarrow (p/q)^2 - 2(p/q) + c = 0$$
$$p/q = \frac{2 \pm \sqrt{(4-4c)}}{2} = 1 \pm \sqrt{(1-c)}$$

$$\text{Now } p \text{ and } q \in R_{n-1} \Rightarrow p/q \in R_{n-1}$$
$$\Rightarrow 1 \pm \sqrt{(1-c)} \in R_{n-1}$$

If we can choose c to be greater than 1, this is clearly a contradiction, since in this case, $\sqrt{(1-c)}$ is not a real number. The following question shows that we may essentially choose c to be of any magnitude.

Q 3.12 Prove that $R_{n-1}(\sqrt{c}) = R_{n-1}(a\sqrt{c})$ where $a \in R_{n-1}$. (i.e. Prove that the field obtained by adjoining $\sqrt{c}$ to a field is the same as that obtained by adjoining a multiple of $\sqrt{c}$ to that field)

Q 3.13 Prove that $\sqrt[3]{4}$ is not an element of any square root extension field of the rationals.

We can conclude from this that no square root extension field of the rationals is 'big' enough to cover the whole number axis. Hence, no matter how far we go in these extensions, we shall still be left with gaps. Of course, by taking cube root extensions, and fourth root extensions, etc. we can plug many of these gaps. Again, unfortunately, not all. Any element of such a sequence of extensions will be of the form

$$\alpha = a + b\sqrt[n]{c}$$

where a, b and $c \in R_{n-1}$ and $\sqrt[n]{c} \notin R_{n-1}$.

$$\Rightarrow (\alpha - a)^n = b^n c$$

The l.h.s. will yield, on expanding, a polynomial of degree n for α, whose coefficients are in R_{n-1}. By a similar sequence of power expansions, we can reduce these coefficients to rational numbers. For instance, let

$$\alpha = 1 + 2\sqrt[3]{4}$$

Then $$(\alpha - 1)^3 = 2^3.4$$

$$\Rightarrow \alpha^3 + 3\alpha - 3\alpha^2 - 1 = 32$$

and so $$\alpha^3 - 3\alpha^2 + 3\alpha - 33 = 0$$

In other words, any element of an extension field of the rationals satisfies a polynomial equation with rational coefficients, and therefore an equation with integer coefficients (multiply through by the highest common factor, to remove fractions).

Hence the numbers we cannot reach by such fields must not occur as the roots of such equations. We call such numbers *non- algebraic* or *transcendental.*

The theory we have developed shows the complexity of the number system. It also shows that one has to be very careful when discussing the roots of certain equations. For instance, the equation,

$$2x^2 - x - 3 = 0$$

has *no* roots in the natural numbers, *one* root in the integers, and *two* roots in the rational field. The equation

$$x^2 + 1 = 0$$

however, has no roots at all in the real number system.

We conclude this chapter with an easy application of our field extension theory—in later chapters we shall use this theorem to establish several other important results.

If a cubic equation with rational coefficients has no rational roots, then none of its roots lies in any of the square root extension fields of the rationals.

The proof of this theorem is relatively simple, but rather long. It breaks down quite naturally into three separate parts, which we now proceed to give. However, we shall prove thc theorem only for a simple case, and then sketch the proof in the general case.

(a) Consider the cubic equation,

$$x^3 + ax^2 + bx + c = 0$$

Each root of this equation arises out of a linear factor of this expression, and so if x_1, x_2, and x_3 are its roots,

$$(x - x_1)(x - x_2)(x - x_3) \quad \text{and} \quad x^3 + ax^2 + bx + c$$

must be different expressions of the same thing. Expanding the l.h.s. gives

$x^3 - (x_1 + x_2 + x_3)x^2 + (x_1x_2 + x_1x_3 + x_2x_3)x - x_1x_2x_3$ is the same

as $x^3 + ax^2 + bx + c$

from which we may deduce

$$x_1 + x_2 + x_3 = -a$$

i.e. We may express the *sum* of the roots of a cubic equation in terms of its coefficients.

(b) Consider the particular equation,

$$x^3 - 6 = 0$$

Now, one root of this equation is $\sqrt[3]{6}$ and this is not rational. Suppose then, that a root x_1 of this equation is an element of one of the square root extension fields of the rationals, i.e. R_n, and further, that this is the 'smallest' such field. Then we have,

$$x_1 = p + q\sqrt{c}$$

where p, q, and $c \in R_{n-1}$ and $\sqrt{c} \notin R_{n-1}$ but $\sqrt{c} \in R_n$
Consider another element of R_n, x_2 where

$$x_2 = p - q\sqrt{c}$$

It is now easy to verify that

$$x_2^3 = p^3 + 3pq^2c - (3p^2q + cq^3)\sqrt{c} \qquad (1)$$

However, $x_1 = p + q\sqrt{2}$ satisfies our original equation, and so:

$$(p + q\sqrt{c})^3 - 6 = 0$$

and expanding this gives

$$p^3 + 3pq^2c + (3p^2q + cq^3)\sqrt{c} = 6 \qquad (2)$$

Equation (2) can be re-arranged to give,

$$\sqrt{c} = \frac{6 - p^3 - 3pq^2c}{3p^2q + cq^3}$$

which implies that $\sqrt{c} \in R_{n-1}$ since p, q, and c are. However, this is false, and so we conclude that we have divided by zero.

$$3p^2q + cq^3 = 0 \qquad (A)$$

Putting this result into equation (2) yields

$$p^3 + 3pq^2c = 6 \qquad (B)$$

If we now put both results (A) and (B) into (1), we have

$$x_2^3 = 6 - (0)\sqrt{c}$$

i.e. $$x_2^3 = 6$$

and so x_2 is another root of the equation. We can further show that these two roots are different, because if

$$x_1 = x_2$$
$$p + q\sqrt{c} = p - q\sqrt{c}$$
$$\Rightarrow \quad 2q\sqrt{c} = 0 \Rightarrow q = 0 \quad \text{or} \quad \sqrt{c} = 0$$

If $q=0$, then $x_1 = p \in R_{n-1}$, and the equation is soluble in the smaller field R_{n-1}, contrary to assumption. Similarly, if $\sqrt{c}=0$ the equation is soluble in the rationals, and we know this to be impossible. Hence, we conclude,

$$x_1 \neq x_2$$

(c) Suppose that a third root of the equation

$$x^3 = 6$$

is x_3.
Then by (a),

$$x_1 + x_2 + x_3 = 0$$
$$(p+q\sqrt{c}) + (p - q\sqrt{c}) + x_3 = 0$$
$$x_3 = -2p$$

However, $p \in R_{n-1}$ and so there is a root in the smaller field R_{n-1}. This contradicts our assumptions, and so there is no smallest field containing these roots, and hence no square root field extension of the rationals containing them.

To prove the theorem generally is no more difficult than the algebra involved. We can obviously omit part (i) since it is already perfectly general.

(b) Consider the equation,

$$x^3 + ax^2 + bx + c = 0$$

Suppose that this has no rational roots, but that it has a root x_1 in some square root extension field R_n, and that this is the smallest such field. Then

$$x_1 = p + q\sqrt{c}$$

where p, q, and $c \in R_{n-1}$ and $\sqrt{c} \notin R_{n-1}$ but $\sqrt{c} \in R_n$
We then show that

$$x_2 = p - q\sqrt{c}$$

is also a root, and that

$$x_1 \neq x_2$$

(c) If the third root is x_3, we have by (i)

$$x_1 + x_2 + x_3 = -a$$

i.e.

$$x_3 = -a - (x_1 + x_2)$$
$$= -a - 2p$$

Hence $x_3 \in R_{n-1}$ and the equation is soluble in a smaller field. This contradicts our assumption, and so no such field can exist.

Further investigations

1 R_0 is the field of rational numbers, and a and $b \in R_0$. A new system is defined on pairs of these elements

$$(a, b)$$

with the following rules:
(a) *Equality* $(a, b) = (c, d)$ whenever $a = c$ and $b = d$
(b) *Addition* $(a, b) + (c, d) = (a + c, b + d)$
(c) *Multiplication* $(a, b).(c, d) = (ac + 3bd, ad + bc)$
What is this new system?
Show that in this system the equation $x^2 - 3 = 0$ is soluble.

2 Find two irrational numbers (both positive) such that their sum is rational (*hint*: consider decimals).

3 Prove that if $a > 1$ and $a \in N$ then $\sqrt{(a^2 - 1)}$ is irrational.

4 R_0 is the field of rational numbers. A new system is defined of pairs of these numbers, with the following rules:
(a) *Equality* $(a, b) = (c, d)$ whenever $a = c$ and $b = d$
(b) *Addition* $(a, b) + (c, d) = (a + c, b + d)$
(c) *Multiplication* $(a, b)(c, d) = (ac - bd, ad + bc)$. Show that in this system, the equation $x^2 + 1 = 0$ is soluble, and that every quadratic equation is also. What is $\sqrt{-1}$?

5 Prove that if $\sqrt[3]{a}$ is not rational and a is rational, then it is not an element of any square root extension of the rationals.

6 Show that the number $a = \sqrt{2} + 2\sqrt[3]{4}$ satisfies a polynomial equation with integer coefficients.

4 The winding function

4.1 Construction of the function

Many situations which are of interest to the mathematician are periodic in nature, exhibiting a repetitive behaviour like the rising and setting of the sun, the tides, and the motion of a pendulum. In order to describe these sorts of situations we need to construct a series of functions which are themselves periodic. We begin to do so by an apparently simple method—winding some string onto a cylinder. However, before long we find ourselves with some surprising results.

Consider a line L representing a fine string and a circle representing a disc (Figure 4.1). The string can be wound tightly round the disc by bringing a point A on the string into contact with a point B on the disc. The string above A can then be wound anticlockwise onto the disc, while the string below can be wound on clockwise.

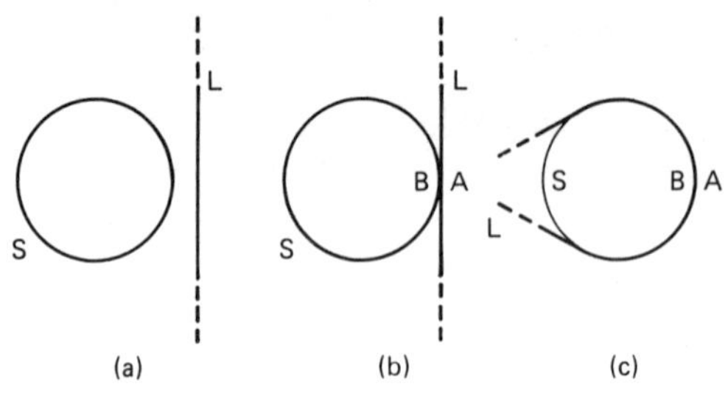

Fig. 4.1 Mapping of points on line L onto circumference S

What this achieves is a mapping of each point of the line onto a point of the circumference of S. Each point of L corresponds to only one point of S. We thus have a function: $L \longrightarrow S$. We shall call this function the *winding function*. Having established the general idea, let us introduce some details and see what we can make of them.

Firstly, let the radius of the circle S be 1 unit, so that the circumference is 2π units, and secondly imagine that the line L extends indefinitely in either direction. Introduce two perpendicular axes, whose origin is at the centre of the disc, and give L the same scale as these axes. L and S can now be written,

$$L = \{k : k \text{ is a real number}\}$$

and

$$S = \{(x, y) : x^2 + y^2 = 1\}$$

If we denote the winding function by W, then obviously we have

$$W : R \longrightarrow R^2$$

As the string is wrapped round the disc in both directions many points of L map onto the same point of S. Clearly, if X and Y are two points of L separated by the circumference of S (i.e. 2π units), they will be wound onto the same point. Hence we have,

$$W(x) = W(y)$$

Similarly, if X and Y are two points of L separated by any multiple of the circumference, they will be wound onto the same point. We can write,

$$W(x) = W(y) \Leftrightarrow x - y = 2n\pi \text{ for } n \text{ an integer.}$$

This can be written in the form,

$$W(x) = W(x + 2n) \text{ for all } n \in J \qquad (1)$$

This is the periodic nature of W—every 2π it repeats itself, and we say that W has period 2π.

We can also obtain values of W for some isolated points of L. Let the point of intersection of L and S be O. (We shall in future always refer to this point as O and *not* mean the origin of the x and y axes.) Let A, B, and C be the points where S intersects the axes as shown in Figure 4.2. Now, OA is a quarter of the circumference of S, and so the arc length OA is,

$$\overset{\frown}{OA} = 2\pi/4 = \pi/2$$

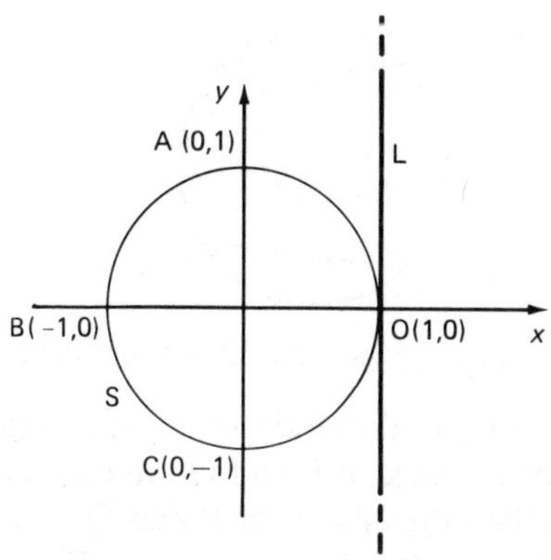

Fig. 4.2 S intersects the axes at A, B, and C

Hence,

$$W(\pi/2) = (0, 1)$$

By exactly similar arguments, we find:

$$W(\pi) = (-1, 0) \quad \text{and} \quad W(3\pi/2) = (0, -1)$$

Thus to find the point on S represented by $W(33\pi/2)$ we proceed as follows:

$$33\pi/2 = 16\pi + \pi/2 = 8(2\pi) + \pi/2$$

Hence
$$\begin{aligned} W(33\pi/2) &= W(8(2\pi) + \pi/2) \\ &= W(\pi/2) \text{ by (1)} \\ &= (0, 1) \text{ by the results obtained above.} \end{aligned}$$

In words, the point on L represented by $33\pi/2$ would have to be wrapped round the disc $8\frac{1}{4}$ times.

We can establish other properties of W by using some of the properties of a circle. Let P and Q be two points on L separated by a distance of π. Then P and Q must be mapped onto diametrically opposite points. Thus if we have,

$$W(p) = (a, b) \quad \text{then} \quad W(p + \pi) = (-a, -b)$$

Q 4.1 If K is a point on L and $W(k) = (a, b)$ what is $W(-k)$?

One useful property of a circle worth taking note of is the equal chord property; equal chords of a circle subtend equal arcs, and equal arcs subtend equal chords (Figure 4.3). If we denote the length of the chord AB by $\overline{AB}$, and the length of the arc cut off by the points A and B by $\widehat{AB}$, we have

$$\overline{AB} = \overline{CD} \Leftrightarrow \widehat{AB} = \widehat{CD}$$

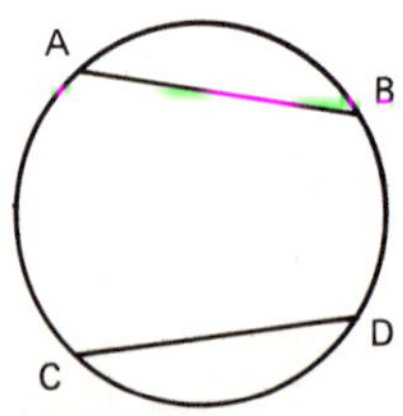

Fig. 4.3 Equal chords of a circle subtend equal arcs

This fundamental property of the circle allows us to calculate many points of the function W. For instance, if P and Q are two points of L separated by a distance of $\pi/2$, then the images of P and Q, P′ and Q′, will also be separated by an arc length of $\pi/2$. However, on Figure 4.4, the points A and O are separated by an arc length of $\pi/2$. By the circle property, they

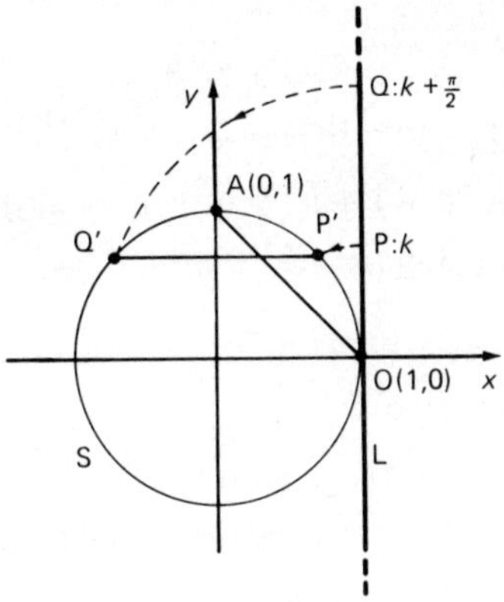

Fig. 4.4 If P and Q are separated by a distance $\pi/2$, so are the image P′ and Q′

subtend equal chords:

$$\overline{P'Q'} = \overline{AO}$$

Let $P' = (a, b)$ and $Q' = (u, v)$ then, this means

$$\sqrt{\{(u-a)^2 + (v-b)^2\}} = \sqrt{\{(0-1)^2 + (1-0)^2\}}$$

$$\Rightarrow \quad u^2 - 2ua + a^2 + v^2 - 2vb + b^2 = 2$$

However, $u^2 + v^2 = 1$ and $a^2 + b^2 = 1$, since P′ and Q′ lie on S,

$$\Rightarrow \quad -2ua - 2vb + 2 = 2$$

i.e. $$ua + vb = 0 \qquad \text{(A)}$$

In addition (Figure 4.5), the distance between $(k + \pi/2)$ and $\pi/2$ on L is the same as the distance between k and O on L, because

$$(k + \pi/2) - \pi/2 = k - 0 = k$$

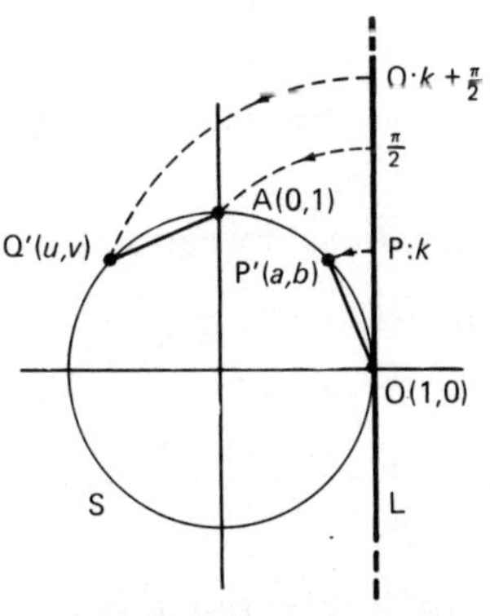

Fig. 4.5 The distance between $(k + \pi/2)$ and $\pi/2$ on L, is the same as the distance between k and O on L

Hence,

$$\overline{P'O} = \overline{Q'A}$$

$$\sqrt{\{(a-1)^2 + (b-0)^2\}} = \sqrt{\{(u-0)^2 + (v-1)^2\}}$$

$$\Rightarrow \quad a^2 - 2a + 1 + b^2 = u^2 + v^2 - 2v + 1$$

When we simplify this using $a^2 + b^2 = u^2 + v^2 = 1$, we have

$$v = a \qquad \text{(B)}$$

Putting (B) into (A) we have

$$ua + ab = 0,$$

$$\Rightarrow \quad u = -b$$

Hence, we have the following result:

If $W(k) = (a, b)$ then $W(k + \pi/2) = (-b, a)$

Q 4.2 Given that $W(k) = (a, b)$ find the value of $W(k + 3\pi/2)$.

Using the same method, we can calculate more points of W. Consider $W(\pi/4)$. Figure 4.6 makes clear that the distance between $\pi/2$ and $\pi/4$ is the same as that between $\pi/4$ and O. Hence they subtend equal arcs on S.

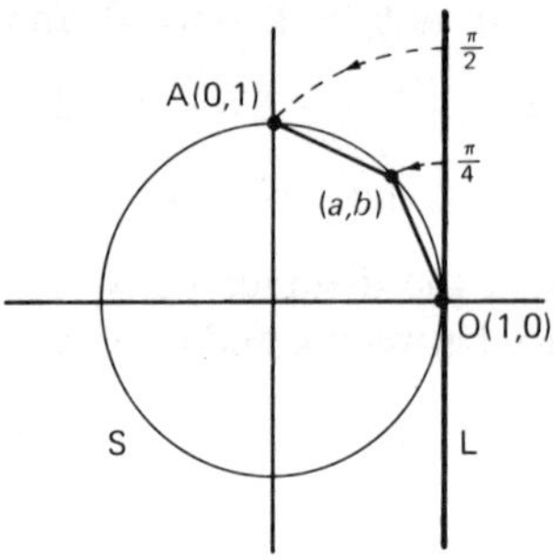

Fig. 4.6 The points subtend equal arcs on S

Because of this, they are connected by equal chords:

$$\Rightarrow \quad \sqrt{\{(0-a)^2 + (1-b)^2\}} = \sqrt{\{(a-1)^2 + (b-0)^2\}}$$

$$\Rightarrow \quad a^2 + 1 - 2b + b^2 = a^2 - 2a + 1 + b^2$$

$$\Rightarrow \quad a = b$$

But $(a, b) \in$ S and so, $a^2 + b^2 = 1$

$$a^2 + a^2 = 1 \Rightarrow a^2 = \tfrac{1}{2}$$

Hence, $$W(\pi/4) = (1/\sqrt{2}, 1/\sqrt{2})$$

(N.B. Of course, we could take the negative square root but it is clear that it is not applicable here.)

Q 4.3 Use this last result to determine $W(\pi/8)$.

Q 4.4 (i) Show that the distance between $\pi/2$ and $\pi/3$ on L is the same as that between $\pi/6$ and O. Hence, show that if $W(\pi/6) = (a, b)$ then $W(\pi/3) = (b, a)$.

(ii) Show additionally, that the distance on L between $\pi/6$ and O is the same as that between $\pi/3$ and $\pi/6$. Hence, find $W(\pi/6)$ and $W(\pi/3)$.

One can continue the method of the previous examples indefinitely, to obtain isolated values for W. However, the following more general approach attacks the same problem in a more systematic manner.

Suppose that K and M are two points of L, and are represented by k and m respectively. If the images of K and M under W are K′ and M′ and P is the point on L represented by $k - m$, we have the situation shown in Figure 4.7.

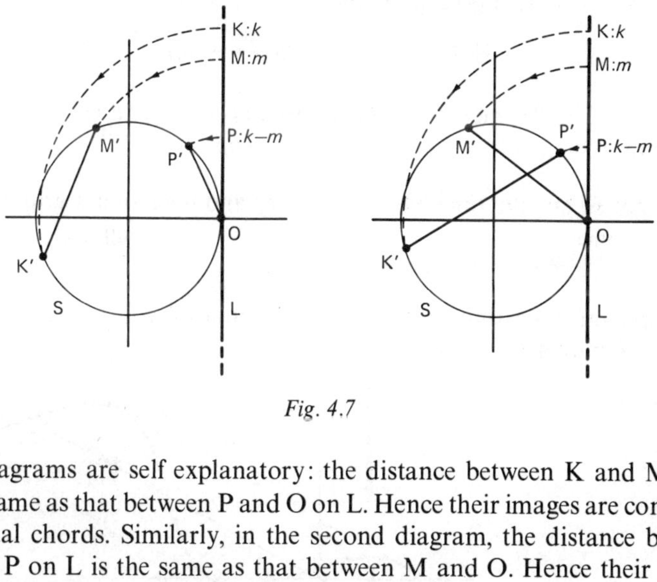

Fig. 4.7

The diagrams are self explanatory: the distance between K and M on L is the same as that between P and O on L. Hence their images are connected by equal chords. Similarly, in the second diagram, the distance between K and P on L is the same as that between M and O. Hence their images are connected by equal chords.

Thus if $W(k) = (a, b)$, $W(m) = (p, q)$ and $W(k - m) = (x, y)$:

$$\sqrt{\{(a-p)^2 + (b-q)^2\}} = \sqrt{\{(x-1)^2 + (y-0)^2\}}$$

$$\Rightarrow \quad a^2 - 2ap + p^2 + b^2 - 2bq + q^2 = x^2 - 2x + 1 + y^2$$

But $(a,b), (p,q)$, and $(x,y) \in S$ and so $a^2 + b^2 = 1$, etc.

$$\Rightarrow \quad -2ap - 2bq = -2x$$

$$\Rightarrow \quad x = ap + bq \qquad \text{(A)}$$

Using the second diagram

$$\sqrt{\{(a-x)^2 + (b-y)^2\}} = \sqrt{\{(p-1)^2 + (q-0)^2\}}$$

$$\Rightarrow \quad a^2 - 2ax + x^2 + b^2 - 2by + y^2 = p^2 - 2p + 1 + q^2$$

Again, $a^2 + b^2 = 1$, etc. this reduces to

$$\begin{aligned} -2ax - 2by &= -2p \\ y &= (p - ax)/b \\ &= (p - a^2p - abq)/b \text{ (using (A))} \\ &= \{p(1-a^2) - abq\}/b \\ &= (pb^2 - abq)/b \\ &= pb - aq \end{aligned}$$

Hence, we have the result

If $W(k) = (a,b)$ and $W(m) = (p,q)$ then

$$W(k-m) = (ap + bq, bp - aq)$$

Q 4.5 Use a similar method to that above to prove that if $W(k) = (a,b)$ and $W(m) = (p,q)$ then $W(k+m) = (ap - bq, aq + bp)$

Having got so far, the time has come to try and construct a graph for W. However, since,

$$W : R \longrightarrow R^2$$

our graph must be three-dimensional. A suitably constructed graph is shown in Figure 4.8.

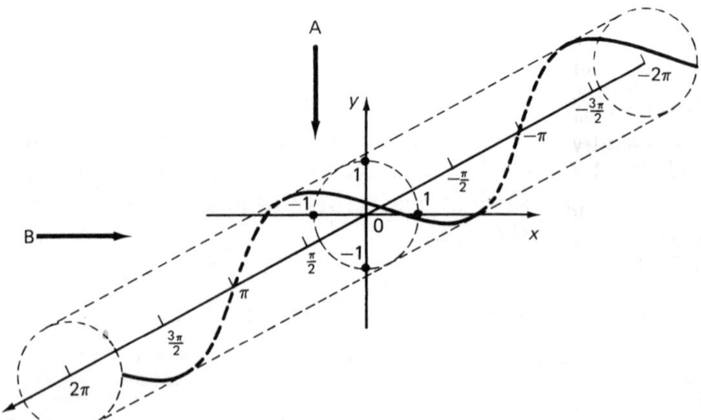

Fig. 4.8 Construction of graph for winding function

Of course only a portion of the graph is shown, but it is not difficult to see that the graph can be extended indefinitely in either direction along the l-axis.

Can we find a formula for W? Since W is a function which associates with a point of L a point on S, we have the correspondence

$$W(r) = (s, t)$$

That is the *number* r is mapped to the numbers s and t. Hence, we can write W in the form

$$W(x) = (f(x), g(x))$$

Unfortunately, however, it is impossible to express these *component functions* f and g in ordinary algebraic terms. For this reason W is called a non-algebraic or *transcendental* function, to distinguish it from functions which are so expressible. It turns out in mathematics, that those functions which are of most significance are all transcendental. Since these functions are important, but we cannot express them algebraically, what is done is to establish some of their properties without recourse to a formula.

Consider the real function f which has the properties:

$$\text{(i)}\ f(x+y) = f(x) + f(y)$$

and

$$\text{(ii)}\quad f(kx) = kf(x)$$

Notice that we have been given only properties of the function, *not* a formula for it. However, these properties can go a long way as we shall see.

Consider $f(0)$. Now, $0 = 0 + 0$

$$\begin{aligned} \Rightarrow\quad f(0) &= f(0+0) \\ &= f(0) + f(0)\ (by\ \text{property (i)}) \\ &= 2f(0) \\ \Rightarrow\quad f(0) &= 0. \end{aligned}$$

Consider $f(x)$. Now, $x = 1.x$

$$\begin{aligned} \Rightarrow\quad f(x) &= f(1.x) \\ &= xf(1)\ \text{(by property (ii))} \end{aligned}$$

This says that $f(x)$ for any x is determined once we know the value of $f(1)$. Now $f(1)$ is constant, say a, so that

$$f(x) = ax$$

Hence, f is 'the straight line through the origin of gradient a'.

Thus while we may not know of a formula for a function, we can very often do without it. In the case of transcendental functions, these sorts of properties must be deduced from the nature of the function itself. This is precisely what we have achieved for the winding function.

Q 4.6 Show that if $W(k) = (a, b)$ then $W(2k) = (a^2 - b^2, 2ab)$.

4.2 Applying the winding function

This last question that you have answered might have left you with a feeling of 'having seen it all before'. Perhaps you feel that the working is familiar but you can't put your finger on it. Well, this feeling was justified. As we remarked earlier, the form of W is such that,

$$W(x) = \{f(x), g(x)\}$$

and we can now reveal that

$$f(x) = \cos x \quad \text{and} \quad g(x) = \sin x$$

If you look at the graph of the winding function again, and imagine the views that you would see in the directions of the arrows A and B you obtain Figure 4.9.

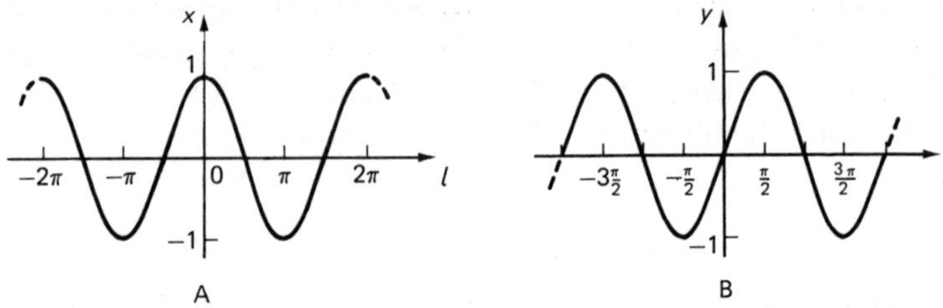

Fig. 4.9 Views of winding function in directions A and B

Is any more convincing necessary? The reason that we did not develop the trigonometric functions from triangles and angles, is that it begs every mathematical question in the book. The simple question, 'What is an angle' could occupy the whole of this book without being satisfactorily answered. The winding function avoids all the unnecessary concepts and produces straight away a function—not a *ratio*, and it does not get bogged down in the details of geometry. That awkward transition from angles to radians (i.e. real numbers) is completely circumvented, and the development of a tool for use in analysis is created immediately and elegantly.

If this isn't reason enough, the following applications of an idea associated with the winding function ought to make it so. That it is applicable in the theory of numbers, is as surprising as it is delightful. It points once more to the power of this approach. As an introduction to this problem consider the following. The real number

$$\sqrt{2}$$

is an irrational number, and it can be represented in the decimal system only by means of orderings. Thus,

$$1.41 < \sqrt{2} < 1.42$$

The problem with irrational numbers like this is finding rational numbers sufficiently accurate to approximate them.

Q 4.7 Prove that $\sqrt{2}$ is an irrational number.

We would like to know: can we find a rational approximation to $\sqrt{2}$ that is as close as we please? The answer is obviously yes, but the proof remains apparently formidable. We shall prove this in a few lines, using a related function.

Consider the sequence,

$$\sqrt{2}, 2\sqrt{2}, 3\sqrt{2}, 4\sqrt{2}, \ldots, (n-1)\sqrt{2}$$

Mark each of these numbers on a number line, and on a similar line mark the numbers,

$$1/n, 2/n, 3/n, 4/n, \ldots, (n-1)/n$$

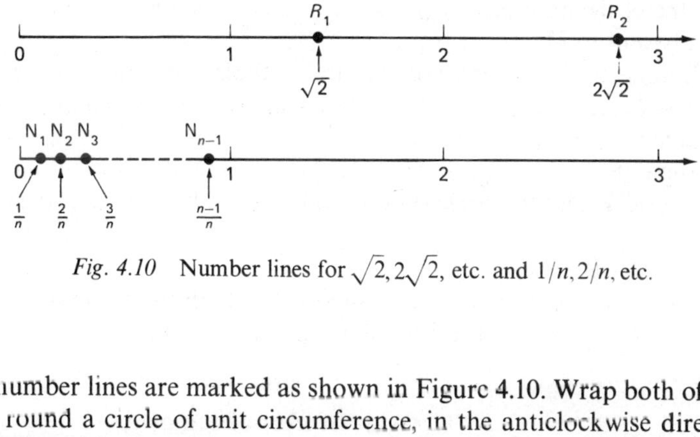

Fig. 4.10 Number lines for $\sqrt{2}, 2\sqrt{2}$, etc. and $1/n, 2/n$, etc.

The number lines are marked as shown in Figure 4.10. Wrap both of these lines round a circle of unit circumference, in the anticlockwise direction, beginning with the origin O at the same point in each case. Because this circle has a unit circumference, each of the numbers

$$0, 1, 2, 3, \ldots$$

is mapped onto the initial point O. The points,

$$N_1, N_2, N_3, N_4, \ldots, N_{n-1}$$

are equally spaced round the circumference of the circle, and hence determine, n equal arcs. The numbers,

$$\sqrt{2}, 2\sqrt{2}, 3\sqrt{2}, \ldots, (n-1)\sqrt{2}$$

determine $n-1$ points R on the circle in Figure 4.11.

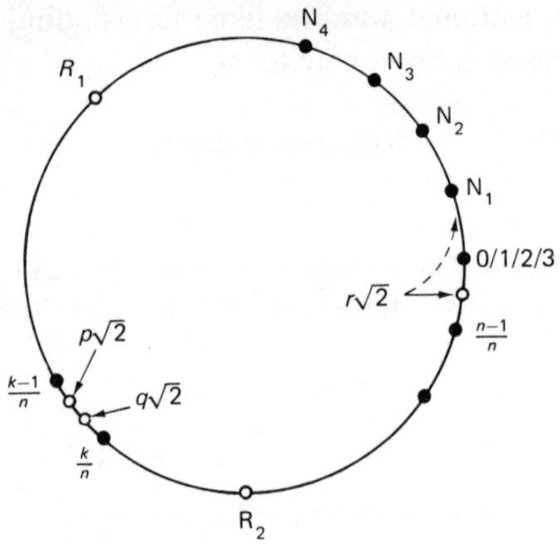

Fig. 4.11 Wrapping both of the number lines in Figure 4.10 round a unit circle

There are two possibilities to consider:

(i) One of the multiples of $\sqrt{2}$, say $r\sqrt{2}$, lands in one of the equal arcs adjacent to O.

(ii) If none of the multiples of $\sqrt{2}$ land in these two arcs, there are then $(n-2)$ arcs for the $(n-1)$ multiples to land in, and so at least two of them must land in the same arc; say $p\sqrt{2}$ and $q\sqrt{2}$.

(i) Assume that $r\sqrt{2}$ lands in an arc adjacent to O. If this point has been 'wrapped' round the circle m times before it lands in this arc, then:

$$r\sqrt{2} - m < 1/n$$

because the distance of $r\sqrt{2}$ from O is less than the arc length $1/n$. Hence, there exist natural numbers, m and r so that

$$\sqrt{2} < m/r + 1/nr \quad \text{with} \quad r \leq n-1$$

$$\Rightarrow \quad \sqrt{2} \simeq m/r \text{ with an error of at most } 1/n$$

(ii) In this second possibility, if $p\sqrt{2}$ and $q\sqrt{2}$ are mapped into the same arc, then the difference,

$$q\sqrt{2} - p\sqrt{2}$$

differs from a whole number of circumferences by an amount less than $1/n$. That is, there is a natural number t such that

$$(q\sqrt{2} - p\sqrt{2}) - t < 1/n$$

$$\Rightarrow \quad (q-p)\sqrt{2} \quad < t + 1/n$$

$$\sqrt{2} \quad < t/(q-p) + 1/n(q-p)$$

$$\Rightarrow \quad \sqrt{2} \simeq t/(q-p) \text{ with an error of at most } 1/n.$$

What we have proved is the following: if we want a rational approximation of $\sqrt{2}$ to be accurate to $1/1\,000\,000$ we shall find a multiple of $\sqrt{2}$ which differs from a natural number by less than $1/1\,000\,000$ in the set

$$\{\sqrt{2}, 2\sqrt{2}, 3\sqrt{2}, \ldots, 999\,999\sqrt{2},\}$$

and hence our approximation.

Q 4.8 Prove the following extension of this; if x is any positive real number, then one of the set

$$\{x, 2x, 3x, 4x, \ldots (n-1)x\}$$

differs from a natural number by at most $1/n$.

We can use this last result to prove a theorem about powers of numbers. Is there a power of 2 beginning with 5? To find such a number, we might proceed as follows:

$$2^1 = 2, 2^2 = 4, 2^3 = 8, 2^4 = 16, 2^5 = 32, 2^6 = 64, \ldots$$

We cannot ever be sure of finding a number in this sequence which begins with 5. However, we can prove that such a number exists, though we might not have the patience to continue the sequence to find it. Consider the set,

$$\{(5, 6), (50, 60), (500, 600), (5000, 6000), (50\,000, 60\,000), \ldots\}$$
$$= \{(5.10^0, 6.10^0), (5.10^1, 6.10^1), (5.10^2, 6.10^2), \ldots\}$$

If we take the logarithms (to the base 10) of each element of this set, we can write it

$$\{(\log 5, \log 6), (1 + \log 5, 1 + \log 6), (2 + \log 5, 2 + \log 6), \ldots\}$$

These numbers are shown in Figure 4.12 on a number line.

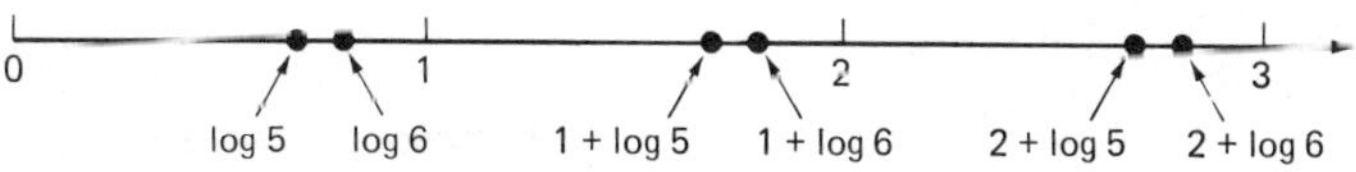

Fig. 4.12 Number line for log 5, log 6, (1 + log 5, 1 + log 6), etc.

If we map this number line onto a circle of unit circumferencc, several interesting things happen:

(i) Clearly, all the numbers $0, 1, 2, 3, \ldots$ are all mapped onto the initial point, which we shall call O.

(ii) Since the difference between successive points of the form $n + \log 5$ is: $(n + \log 5) - (n - 1 + \log 5) = 1$ each of these points is mapped onto a single point, called A. Similarly, every point of the form $n + \log 6$, is mapped onto a single point B (Figure 4.13).

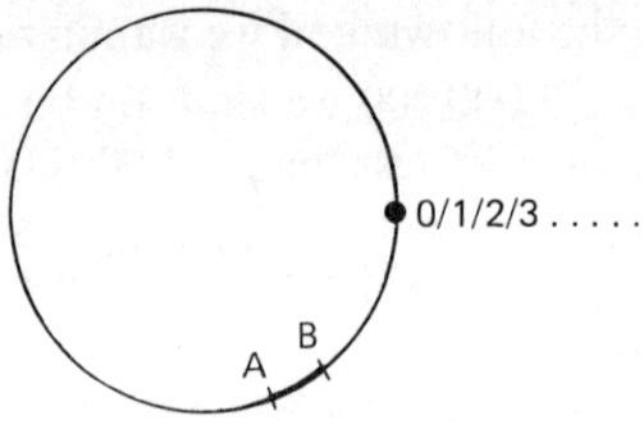

Fig. 4.13 Points are mapped onto A and B only

Next, consider the set:

$$\{2, 2^2, 2^3, 2^4, 2^5, \ldots\}$$

If we take logarithms (to the base 10) of each element of this set we have

$$\{\log 2, 2\log 2, 3\log 2, 4\log 2, 5\log 2, \ldots\}$$

These numbers plotted on a number line are shown in Figure 4.14.

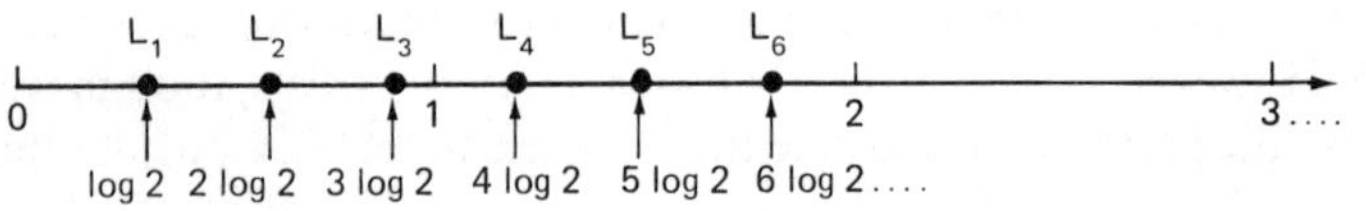

Fig. 4.14 Number line for log 2, 2 log 2, 3 log 2, etc.

These points are marked, $L_1, L_2, \ldots$, etc.
If we map this number line onto the same circle we have Figure 4.15.

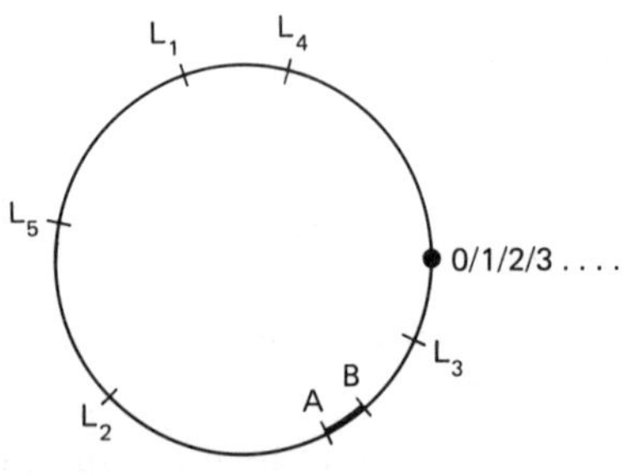

Fig. 4.15 Mapping the number line in Figure 4.14 onto a circle

There are two important properties of this mapping now:

(i) No two of the L's are mapped onto the same point. For, suppose that L_i and L_j are mapped onto the same point, then they differ by a certain number of circumferences.

$$\Rightarrow \quad L_i - L_j = k, \text{ a natural number}$$
$$\Rightarrow \quad i \log 2 - j \log 2 = k$$
$$\Rightarrow \quad \log 2 = \frac{k}{i-j}$$

Hence, log 2 is rational. However, log 2 is not rational, because if

$$\log 2 = a/b \Rightarrow 2 = 10^{a/b} \Rightarrow 2^b = 10^a$$

This is impossible because the right hand side is divisible by 5 while the left is not. And so the L's are distinct.

(ii) Using the result of Question 4.8 one of the set

$$\{\log 2, 2\log 2, 3\log 2, 4\log 2, \ldots\}$$

is as close to a natural number as we please. Let $p \log 2$ be different from a natural number by less than $(\log 6 - \log 5)$. This means that L_p is closer to 0 than A is to B (Figure 4.16). Notice that we don't know which side of 0 L_p is on.

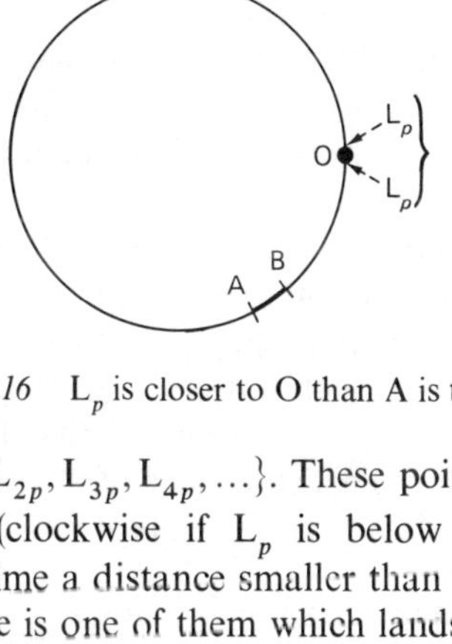

Fig. 4.16 L_p is closer to O than A is to B

Consider the set, $\{L_p, L_{2p}, L_{3p}, L_{4p}, \ldots\}$. These points must march round the circle advancing (clockwise if L_p is below 0, and anticlockwise if L_p is above 0) each time a distance smaller than that between A and B. Hence, eventually, there is one of them which lands up in AB. Let this be the point L_{rp} (Figure 4.17).

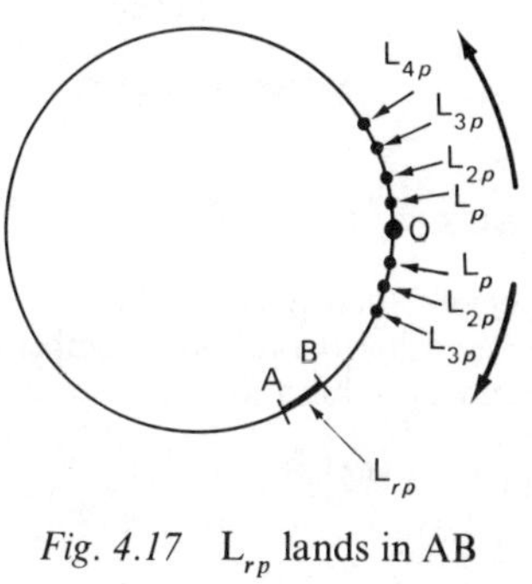

Fig. 4.17 L_{rp} lands in AB

Now A is the image of the points,

$$\{\log 5, 1 + \log 5, 2 + \log 5, \ldots\}$$

and B is the image of the points,

$$\{\log 6, 1 + \log 6, 2 + \log 6, \ldots\}$$

Hence, since L_{rp} lies between A and B, we have

$$A \leq L_{rp} < B$$

(We need an equality here since our 'smaller' steps might take us exactly to A. If it took us exactly to B, we could choose the one before which would be between A and B since the steps are less than AB.)

$$\begin{aligned}
&\Rightarrow & n + \log 5 &\leq rp \log 2 < n + \log 6\\
&\Rightarrow & n \log 10 + \log 5 &\leq rp \log 2 < n \log 10 + \log 6\\
&\Rightarrow & \log 10^n + \log 5 &\leq rp \log 2 < \log 10^n + \log 6\\
&\Rightarrow & \log (5.10^n) &\leq \log 2^{rp} < \log (6.10^n)\\
&\Rightarrow & 5.10^n &\leq 2^{rp} < 6.10^n
\end{aligned}$$

Hence, there is a power of 2 (the *rp*th power) which starts with a 5.

Q 4.9 Show that there is a power of 4 beginning with the digits 34.

As this last question shows, this remarkable theorem can be adapted to show that if a ($\neq 10^n$) is a natural number, then there is a power of a which begins with any sequence of digits we want. However, although the theorem guarantees the existence of such powers, it gives us no clue as to how we should find them. It is worth pointing out that the smallest power of 2 beginning with 9, is 2^{53}! To get some idea of the size of such a number, set

$$\begin{aligned}
x &= 2^{53}\\
\text{then} \quad \log x &= 53 \times \log 2\\
&\simeq 53 \times 0{\cdot}30103\\
&= 15{\cdot}95459\\
\Rightarrow \quad & x \text{ has 16 digits.}
\end{aligned}$$

Further investigations

1 Find the smallest power of 2 that begins with 7.
2 Under what circumstances is $\log_a b$ a rational number?
3 Using the winding function, establish formulae for (i) $\cos 4x$, (ii) $\cos 3x$, and (iii) $\cos 2x$, in terms of $\cos x$.
4 Is 100^{1977} greater than 1977^{100}?

5 Counting without fingers

5.1 The paradoxes of the infinite

Tristram Shandy (the central character in a book of that name) was rather bewildered. He started to write his autobiography, but soon found that he could write about only half a day's experiences in one day of writing. Thus, he reasoned, even if he were to commence writing at birth, and even if he were to live for ever, he could never record his whole life, because at any time he could only have dealt with half his life. But on the other hand, he reasoned, if he were to live for ever, he ought to be able to record his whole life, for the experiences of his first ten years would be recorded by the end of his twentieth year, the experiences of his first twenty years by the end of his fortieth year and so on. Every year of his life would be recorded at some time. Hence, depending on which way he thought about it, he could or couldn't complete his autobiography.

Tristram's inability to resolve this paradox was really to be expected, for his problem involved an infinitude of time: whenever the words *and so on* (...) are used, a claim is made that a progression or sequence of events may be continued indefinitely to a definite conclusion. We say that the entire sequence of events is *infinite* in nature.

Paradoxes of the infinite have been known to mathematics since the time of the Greeks. The examples below, represent a fair sample of these.

Example 1

Consider the 'number' a, defined by,

$$a = \sqrt{2 + \sqrt{2 + \sqrt{2 + \sqrt{2 + \ldots}}}}$$

The dots indicate that to form the quantity a, we must continue this computation indefinitely. It is therefore clear that *a cannot* exist, because however far one proceeds in the calculation, there are still more stages to go. However, let us for a moment assume that *a does exist*.

Then,

$$\begin{aligned} a &= \sqrt{2 + \sqrt{2 + \sqrt{2 + \sqrt{2 + \ldots}}}} \\ &= \sqrt{2 + \sqrt{2 + \sqrt{2 + \sqrt{2 + \ldots}}}} \\ &= \sqrt{2 + a} \end{aligned}$$

$$
\begin{aligned}
&\Rightarrow \quad a^2 = 2 + a \\
&\Rightarrow \quad a^2 - a - 2 = 0 \\
&\Rightarrow \quad (a+1)(a-2) = 0 \\
&\Rightarrow \quad a = -1 \quad \text{or} \quad a = 2
\end{aligned}
$$

Q 5.1 (a) Account for the two 'values' of a.
(b) Evaluate

$$\sqrt{2+\sqrt{2+\sqrt{2+\sqrt{2}}}}$$

Our conclusion is this: if we question the existence of a, then it does not exist; however, if we assume that a exists we can evaluate it.

Example 2
Let us define another 'number', b, by

$$b = \tfrac{1}{2} - \tfrac{1}{2} + \tfrac{1}{2} - \tfrac{1}{2} + \ldots$$

Again, the dots indicate that to form the quantity b, we must continue this computation indefinitely, and therefore b *cannot exist*. However, if we assume that b *does exist*, we may write,

(a)
$$
\begin{aligned}
b &= (\tfrac{1}{2} - \tfrac{1}{2}) + (\tfrac{1}{2} - \tfrac{1}{2}) + (\tfrac{1}{2} - \tfrac{1}{2}) + \ldots \\
&= \quad 0 \quad + \quad 0 \quad + \quad 0 \quad + \ldots \\
&= 0
\end{aligned}
$$

Similarly,

(*b*)
$$
\begin{aligned}
b &= \tfrac{1}{2} - \tfrac{1}{2} + \tfrac{1}{2} - \tfrac{1}{2} + \tfrac{1}{2} - \tfrac{1}{2} + \tfrac{1}{2} - \tfrac{1}{2} + \ldots \\
&= \tfrac{1}{2} + (-\tfrac{1}{2} + \tfrac{1}{2}) + (-\tfrac{1}{2} + \tfrac{1}{2}) + (-\tfrac{1}{2} + \tfrac{1}{2}) + \ldots \\
&= \tfrac{1}{2} + \quad 0 \quad + \quad 0 \quad + \quad 0 \quad + \ldots \\
&= \tfrac{1}{2}
\end{aligned}
$$

Q 5.2 (c) Show that $b = -\tfrac{1}{2}$.

We reach a very different conclusion this time; if we question the existence of b then it does not exist, but if we assume that it exists it may be represented by any of the numbers, 0, $\tfrac{1}{2}$, or $-\tfrac{1}{2}$.

Example 3
Consider the circle in Figure 5.1, and the diameter AB. If the circle has a radius of r, then

$$AB = 2r$$

As a first approximation we may write;

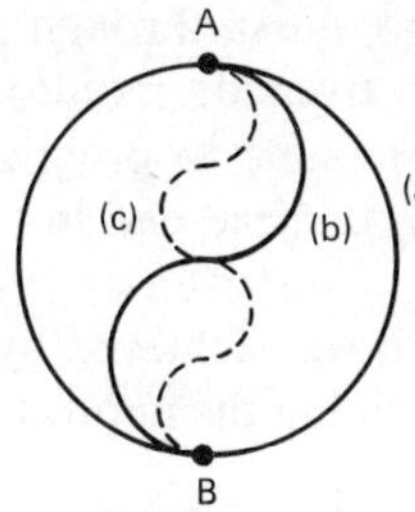

Fig. 5.1

(a) $AB \simeq C/2$ (half the circumference)

$\Rightarrow \quad AB \simeq 2\pi r/2 = \pi r$

As a second and closer approximation we may write;

(b) $AB \simeq (2\pi r/2)/2 + (2\pi r/2)/2$

$\Rightarrow \quad AB \simeq \pi r$

As a third, and even better, approximation we have;

(c) $AB \simeq (2\pi r/4)/2 + (2\pi r/4)/2 + (2\pi r/4)/2 + (2\pi r/4)/2$

$\Rightarrow \quad AB \simeq \pi r$

If we assume that this construction is carried out indefinitely, the approximations become closer and closer to the line AB, and eventually merge with it, and so,

$$AB = \pi r$$

However, $AB = 2r$ and hence $\pi = 2$

These examples all point to the fact that the use of the infinite is fraught with danger, and that in this form it should properly be banished from the realm of mathematics. Most mathematicians from Greek times onwards were at one and the same time attracted and repelled by the idea of the infinite. The problem came to a head during the middle of the nineteenth century when mathematics could no longer dispense with the concept. Before this time, paradoxes of the infinite could safely be looked upon as curiosities, more problems of philosophy than mathematics. However, during the period 1600 to 1850 mathematics made gigantic strides, but these advances had only been made possible by involving the use of the infinite. The mathematics of this period was only as strong as the foundations which supported it, and as our examples show, these were very shaky indeed.

The foundations were not easily re-built. The gaps left by the pioneers were undermined by paradoxes, contradictions, and more paradoxes. There developed an imperative need for critical thinkers with imagination and daring—imagination to burst through the intellectual barriers of centuries, and the daring to override intuition and common sense. In a very real way, it was only possible to progress by abandoning common sense and turning back from its dead end in the hopes of finding another pathway.

The decisive breakthrough was achieved by Georg Cantor. He started with the basic idea that describing the natural numbers,

$$\{1, 2, 3, 4, 5, \ldots\}$$

as an infinite set, poses more problems than it resolves, for all this affirms is that they are not finite. Such a description is as useful as saying that *Agaricus campestris* is not a toadstool. He realized that a positive, and not a negative description of the infinite was required.

Cantor recognized, then, that it was not possible to count an infinite set. But was counting at the heart of numerating a set? Suppose that the auditorium of a hall is full. One does not have to count to realize that there is the *same number* of spectators as seats. In other words, because the set of spectators and the set of seats may be paired one with the other, they must be equally numerous. Cantor referred to this pairing as a *one–one correspondence*, and with this tool the problem of the infinite is resolved. Counting does not begin,

$$1, 2, 3, 4, 5, \ldots$$

but begins by pairing the elements of a given set with the elements of a standard set—the natural numbers. The natural numbers are only the labels we attach to these sets.

Thus Cantor made the following definition.

Two sets A and B can be put into one–one correspondence (*written* $A \sim B$) *if each element of A can be associated with just one element of B, and vice versa. A and B are then equi-numerous and are said to have the same cardinal number, or to be equivalent sets.*

Because this definition makes no use of counting, there is nowhere the implicit assumption that the sets have to be finite. The concept of one–one correspondence and cardinal numbers applies to infinite sets just as well as finite sets. As an example of this, consider the two infinite sets,

$$N = \{1, 2, 3, 4, 5, \ldots\}$$

and

$$\updownarrow\ \updownarrow\ \updownarrow\ \updownarrow\ \updownarrow$$

$$E = \{2, 4, 6, 8, 10, \ldots\}$$

Now clearly N and E can be put into one–one correspondence, as illustrated by the arrows joining corresponding elements. We must conclude that there are as many even numbers as there are natural numbers, and therefore that the cardinal numbers of these sets are the same. On the face of it this would seem to be absurd because obviously the even numbers are contained in the natural numbers, and so there cannot be the 'same' number of them. This objection stems from our experience of finite sets. Clearly, if a set is finite, i.e. if it contains n elements and no more, then it cannot be in one–one correspondence with any of its proper subsets since any proper subset could contain at most $n - 1$ elements. This apparent contradiction, 'The whole is not greater than its parts' shows what surprises are to be expected of the infinite.

Just as all the sets containing a trio of elements are represented by the cardinal number 3, any set which is in one–one correspondence with the natural numbers is represented by the new cardinal number.

$$\aleph_0$$

It is called aleph-nought and is a *transfinite number.*

To say that the number of natural numbers, as well as the number of any set of objects in one–one correspondence with it is $\aleph_0$, does not seem to answer the fundamental question of how many there are in each set. It is tempting to think that $\aleph_0$ has no meaning; if this is the case then no number has any meaning. $\aleph_0$ is just as meaningful as one million or the number three. These numbers mean something to us because we can readily call to mind a group of objects for which these numbers denote their numerosity. Thus for a million we might imagine the population of a large town, for three, a musical trio, and for $\aleph_0$ the set of natural numbers. Just as a child grasps the meaning of three by associating with it three fingers or three toes, so the mathematician may grasp the meaning of $\aleph_0$ by becoming familiar with collections having $\aleph_0$ objects in them. Cantor's theory permits him to decide which these collections are.

Q 5.3 Show that there are $\aleph_0$ square natural numbers.

One of Cantor's first discoveries was that the integers had the same cardinal number as the natural numbers. To prove that this is the case, we need only show that there is a one–one correspondence between the integers and the naturals. This is illustrated below;

$$\begin{array}{l} N = \{1, 2, \quad 3, 4, \quad 5, 6, \ldots\} \\ \quad\;\; \updownarrow\, \updownarrow \quad\;\; \updownarrow\, \updownarrow \quad\;\; \updownarrow\, \updownarrow \\ J = \{0, 1, -1, 2, -2, 3, \ldots\} \end{array}$$

Another result that Cantor obtained was the surprising fact that the rational numbers and the natural numbers have the same cardinal number. This is surprising because at first glance the two sets are radically different—

in Chapter 3 we showed that between any two rational numbers there are an infinite number of others, whereas between certain pairs of natural numbers there are no other natural numbers. However, be this as it may, one can prove that these two sets are equivalent by a diagonal process shown below. First of all, we note that every rational number can be written in the form a/b where a and b are integers. We put these rational numbers into the array shown in Figure 5.2, with a/b in the 'ath' column and 'bth' row:

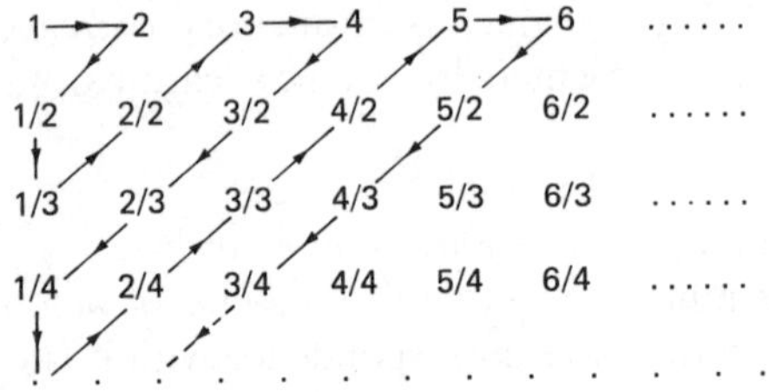

Fig. 5.2

The arrow then gives the following sequence of rational numbers:

$$1, 2, 1/2, 1/3, 2/2, 3, 4, 3/2, 2/3, 1/4, 1/5, 2/4, 3/3, \ldots$$

Of course, many of these rational numbers are repeated several times, so we erase all those numbers a/b for which a and b have a common factor, so that each rational number will appear exactly once and in its simplest form. Thus,

$$1, 2, 1/2, 1/3, 3, 4, 3/2, 2/3, 1/4, 1/5, 5, \ldots$$

and of course, this may be put into one–one correspondence with the natural numbers.

Q 5.4 This shows that the positive rational numbers have the same cardinal number as the naturals. Extend this result to show that the whole set of rational numbers (positive, negative, and zero) are also equi-numerous.

Since the cardinal number of the rationals, the integers, and the natural numbers is $\aleph_0$, it is tempting to suppose that every infinite set has too, and that this is the ultimate result of the analysis of the infinite. Cantor made another significant discovery when he found that the set of all real numbers (R) does not have $\aleph_0$ as its cardinal number. In other words, the totality of real numbers presents a radically different, and so to speak, higher type of infinity than the set of naturals. The proof is very simple. Suppose that the real numbers have the same cardinal number as the naturals. Then they may be arranged in some sequence,

$$1\text{st}, 2\text{nd}, 3\text{rd}, 4\text{th}, 5\text{th}, \ldots$$

Let us write these in decimal form in the following way:

1st	$N_1.a_1a_2a_3a_4a_5a_6\ldots$
2nd	$N_2.b_1b_2b_3b_4b_5b_6\ldots$
3rd	$N_3.c_1c_2c_3c_4c_5c_6\ldots$
	$\ldots$

where the Ns denote the whole part of the number, and the small letters denote the digits after the decimal place. If the real numbers and the naturals have the same cardinal number then this list contains *all* the real numbers. We shall show that this cannot be the case by constructing a new real number which does not belong to this list.

Choose, first of all, a digit a which differs from a_1, and is neither 0 or 9 (to avoid possible ambiguities which may arise from equalities like, $0.9999999\ldots = 1.0000000\ldots$.), then a digit b different from b_2 and again unequal to 0 or 9, similarly c different from c_3 and so on.

Consider the infinite decimal

$$x = 0.abcde\ldots.$$

This new number x is certainly different from any of the numbers in the list above; it cannot be equal to the first because it differs from it in the first digit after the decimal point; it cannot be equal to the second because it differs from it in the second digit after the decimal point; and in general it cannot be identical with the nth number in the list since it differs from it in at least the nth digit after the decimal point. Hence our list does *not* contain all the real numbers, and so the cardinal number of the real numbers is not $\aleph_0$. We usually denote the cardinal number of the real numbers by C (for *continuum*). In fact one may prove that the set of numbers between 0 and 1 have the same cardinal number C. This is easily proved geometrically. The set of real numbers is itself in one–one correspondence with the points on a line. Hence let the points on the line L represent all the real numbers. We construct above this the 'bent unit segment (0, 1)', together with the point A (Figure 5.3).

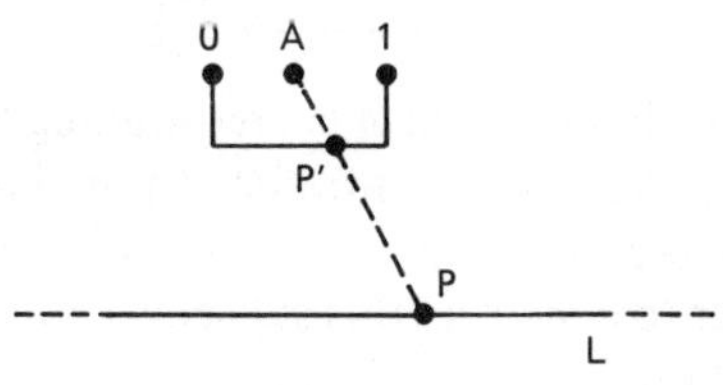

Fig. 5.3

Given any point P on L (and hence any real number) we can associate with it the point P′ (and hence any real number between 0 and 1) by joining P to A. This sets up the required one–one correspondence and the result follows.

Q 5.5 Show that any two segments of the real numbers have the same cardinal number.

Thus far we have identified two 'new' numbers, $\aleph_0$ and C, and suggested that C was 'bigger' than $\aleph_0$. It is intuitively obvious that C must be bigger than $\aleph_0$, but since our common sense is not a reliable guide in the domain of the infinite, we ought to prove this. In order to achieve this we need a very precise definition of what 'bigger' means. Hence we say:

If set A is in one–one correspondence with some subset of set B, while B is not in one–one correspondence with any subset of A, or A itself, then set B has a greater cardinal number than set A.

The set of natural numbers is a subset of the real numbers, while the set of real numbers is not in one–one correspondence with either the natural numbers or any subset of it. Hence according to our definition, the continuum of real numbers has a greater cardinal number than the set of natural numbers.

One question now poses itself; are there other, still larger, transfinite numbers? The answer to this question is very definitely *yes*; in fact there is a transfinite number of them! Cantor again gave a proof that this was the case, but the proof is rather abstract and difficult to follow. Essentially, Cantor proved that starting with any set A, it is possible to construct a set B with a greater cardinal number.

Let A be any set, and define B to be the set of subsets of A. By the word subset, we shall include not only the proper subsets of A, but also the set A itself, and the empty set containing no elements. (Thus if A consists of the three elements, $A = \{a, b, c\}$ then B contains the eight elements, $B = \{\{a,b,c\}, \{a,b\}, \{a,c\}, \{b,c\}, \{a\}, \{b\}, \{c\}, \varnothing\}$.)

We observe first of all that A is a subset of B. Suppose that B can be put into one–one correspondence with A or to some subset of it, i.e. that there is a one–one correspondence between the subsets of A and A or some subset of it,

$$S_a \longleftrightarrow a$$

where we denote by S_a the subset of A corresponding to the element a of A. We shall show that this leads to a contradiction by showing that there is an element of B (i.e. a subset T of A) which cannot have any element corresponding to it. In order to construct this subset we note that for any element x of A, two possibilities exist:

(i) either the set S_x corresponding to x in the one–one correspondence contains x,

or

(ii) it does not.

We now define T to be the subset of A consisting of all those elements x

such that S_x does not contain x. This subset T has remarkable properties for if,

$$a \in S_a \quad \text{then by definition,} \quad a \notin T$$

and conversely, if,

$$a \in T \quad \text{then,} \quad a \notin S_a$$

Obviously then T must be different from every S_a and so the correspondence

$$S_a \longrightarrow a$$

cannot include every subset of A since it does not include T. Hence there is no one–one correspondence between B and A, and so by our definition,

$$\text{cardinal of } A < \text{cardinal of } B$$

Q 5.6 Prove that if a set A has n elements, then the set of subsets of A has 2^n elements.

We denote by $l(A)$ the cardinal number of the set A.

5.2 Prime numbers

The prime numbers are the building blocks from which all other integers may be constructed. Accordingly, the prime numbers have received much study, and considerable efforts have been spent trying to determine the nature of their distribution in the sequence of positive integers. One glance at the list of primes shows the irregularity of their distribution.

$$2, 3, 5, 7, 11, 13, 17, 19, 23, \ldots$$

The dots at the end of this sequence seem to indicate that the primes continue indefinitely. Must this be so, or is there a largest prime? Euclid gave a proof that no such largest prime exists, and the elegance of this proof has been admired by mathematicians of all ages. He proceeded as follows. Suppose that there is a largest prime number p. Then the prime numbers are,

$$2, 3, 5, 7, \ldots, p$$

Consider the number,

$$N = 2.3.5.7 \ldots p + 1$$

This number cannot be divisible by 2, since it would cancel out in the first part of the number and in the second we would be left with a remainder of 1. Similarly, N is not divisible by any of the numbers

$$3, 5, 7, \ldots, p$$

Hence either N is prime, or it is divisible by some prime which is not in our list. Clearly, both N and this other prime are greater than p.

Q 5.7 Prove that any integer greater than 1 is either a prime number or has a prime factor.

The fact that the primes form the building blocks of all other numbers follows from this fact that any natural number is either prime or can be written as the product of prime factors.

When a number has been expressed as a product of primes we may arrange the prime factors in any order. Experience suggests, however, that except for this freedom of order, the decomposition of a number N into primes is unique; i.e. every integer N greater than 1 can be factorized into primes in only one way. This statement seems at first glance so reasonable and so obvious that it might be thought too trivial to bother with proving it. However, a proof of this statement (or better, theorem) is no easy matter and a very subtle argument is needed to establish it. To illustrate the difficulty consider the set of numbers S, each of the form $4n + 1$, where n is a natural number. Then,

$$S = \{5, 9, 13, 17, 21, 25, 29, \ldots\}$$

This set is closed to the operation of multiplication because,

$$\begin{aligned}(4n + 1)(4m + 1) &= 16nm + 4n + 4m + 1 \\ &= 4(4nm + n + m) + 1\end{aligned}$$

which is again of the form $4N + 1$. Note, however, that S is not closed to either of the operations addition and subtraction, since

$$(4n + 1) \pm (4m + 1) = 4(n \pm m) + \begin{cases} 2 \\ 0 \end{cases}$$

neither of which is of the form $4N + 1$.

We define a *pseudo-prime* element of this set as an element which has no other factors in the set. Thus 21 is a pseudo-prime, while 25 is composite.

Within the set S it is true that every element can be factorized. However, it is not true that it can be done uniquely. For instance,

$693 = 4.173 + 1$ is an element of S, while

$693 = 9 \times 77$ or 21×33 all of which are pseudo-primes.

This suggests that the concept of unique factorization amongst the integers is based not only on the operation of multiplication, as it must be, but also on the operation of either addition or subtraction.

Q 5.8 Suppose that a natural number N can be factorized with the primes $p_1, p_2, \ldots$, each raised to the powers $a_1, a_2, \ldots$

i.e. $$N = (p_1^{a_1}).(p_2^{a_2})\ldots(p_r^{a_r})$$

Find an expression for the number of divisors of N.

One way of locating prime numbers is to use a method called the *sieve of Eratosthenes*, named after a mathematician from Alexander who lived about 200 BC. His method proceeds as follows;

Write out the sequence of all integers from 1 onwards to as far as one chooses,

1 2 3 4 5 6 7 8 9 10 11 12 13 14 15 16 17...

We commence with the prime 2. We eliminate every second number after 2 (/):

1 2 3 ~~4~~ 5 ~~6~~ 7 ~~8~~ 9 ~~10~~ 11 ~~12~~ 13 ~~14~~ 15 ~~16~~ 17...

After this has been done, the first unmarked number is 3. It is a prime since it is not divisible by 2. We now mark every third number after 3 (∗):

1 2 3 ~~4~~ 5 ~~6~~∗ 7 ~~8~~ ∗ ~~10~~ 11 ~~12~~∗ 13 ~~14~~ ∗ ~~16~~ 17...

The first number after this is 5. It is prime since it is not divisible either by 2 or 3. We mark every fifth number after 5 (@):

1 2 3 ~~4~~ 5 ~~6~~∗ 7 ~~8~~ ∗ ~~10~~@ 11 ~~12~~∗ 13 ~~14~~ ∗@ ~~16~~ 17...

The first number after this is 7. It is prime since it is not divisible by 2, 3, or 5. One can obviously extend this as far as one pleases. Note that we are using the property established in Question 1: either an integer is a prime, or it is divisible by a prime.

Sooner or later (and most probably later) one can establish whether any given integer is prime or not. However, the method is extremely inefficient. Another, more practicable method, to test a number for primeness is by testing for divisibility with each prime number smaller than it. For instance, suppose we wished to establish whether 97 was prime or not. We could proceed as follows:

97 is not divisible by 2

97 is not divisible by 3

97 is not divisible by 5

...

97 is not divisible by 89

Hence, 97 is a prime number. Can you see that we could shorten this working?

Q 5.9 Given the integer N to test whether it is a prime or not, show that only tests for divisibility of the primes less than or equal to $\sqrt{N}$ need be attempted.

Q 5.10 Using this last result, determine whether 661 and 803 are prime.

A modified version of the sieve of Eratosthenes is much more efficient for filtering out prime numbers. The table below is constructed first;

4	7	10	13	16	19	22	...
7	12	17	22	27	32	37	...
10	17	24	31	38	45	52	...
13	22	31	40	49	58	67	...
16	27	38	49	60	71	82	...
.	.	.	.	.	.	.	
.	.	.	.	.	.	.	

The remarkable property of this table is the following—if N occurs in the table, then $2N + 1$ is not a prime number; and if N does not occur in the table, then $2N + 1$ is a prime number.

In order to establish these properties of the table we shall prove separately each part of the statement. As a first step we need to establish something about the numbers comprising the table. The numbers in the first row are

4	7	10	13	16	19 ...
1st	2nd	3rd	4th	5th	6th ...

which can be re-written

4	$4+3$	$4+2.3$	$4+3.3$	$4+4.3$	...
1	2	3	4	5	

Can you see that this means the nth number in the first row is

$$\begin{aligned} &4+(n-1)3 \\ &= 4+3n-3 \\ &= 1+3n \end{aligned}$$

This is also the first number in the nth row. Now each row increases as follows,

$$\begin{array}{lll} \text{First row} \longrightarrow 3 & (4 \overset{3}{\ } 7 \overset{3}{\ } 10 \overset{3}{\ } 13 \ \ldots) \\ \text{Second row} \longrightarrow 5 & (7 \overset{5}{\ } 12 \overset{5}{\ } 17 \overset{5}{\ } 22 \ \ldots) \\ \text{Third row} \longrightarrow 7 & (10 \overset{7}{\ } 17 \overset{7}{\ } 24 \overset{7}{\ } 31 \ \ldots) \end{array}$$

Can you see that the nth row increases by $2n + 1$? Hence, the nth row looks like

$1 + 3n$	$1 + 3n + 2n + 1$	$1 + 3n + 2(2n + 1)$	...
1st	2nd	3rd	...

This means that the mth number in the nth row is:

$$\begin{aligned} &1 + 3n + (m - 1)(2n + 1) \\ &= 1 + 3n + 2mn + m - 1 - 2n \\ &= n + m + 2mn \\ &= (2m + 1)n + m \qquad (1) \end{aligned}$$

Now, if N occurs in the table, then

$$N = (2m + 1)n + m$$

for some natural numbers n and m. Hence,

$$\begin{aligned} 2N + 1 &= 2((2m + 1)n + m) + 1 \\ &= 2(2m + 1)n + 2m + 1 \\ &= (2m + 1)(2n + 1) \end{aligned}$$

This shows that the factors of $2N + 1$ are $(2m + 1)$ and $(2n + 1)$ and so it cannot be prime. This completes the first part of the proof.

Next, we must show that if N is not in the table, then $2N + 1$ is a prime. We can logically re-arrange this to the following: if $2N + 1$ is not prime, then N is in the table. Suppose then, that $2N + 1$ is not prime. Hence $2N + 1$ has two factors greater than 1, i.e.

$$2N + 1 = a.b$$

Now $2N + 1$ is an odd number (why?). Only the product of two odd numbers is odd, and so a and b must each be odd. Hence, we can write

$$a = 2s + 1 \quad \text{and} \quad b = 2t + 1$$

$2N + 1 = a.b$ becomes,

$$\begin{aligned} 2N + 1 &= (2s + 1)(2t + 1) \\ &= 2s(2t + 1) + 2t + 1 \\ 2N &= 2s(2t + 1) + 2t \\ N &= s(2t + 1) + t \end{aligned}$$

The form of this equation in s and t is exactly the same as the form of the equation (1) in n and m respectively. Hence N occurs in the table as the tth number in the sth row.

The conclusion is that $2N + 1$ is prime if and only if N does not occur in the table.

Q 5.11 Use this table to show that 6 853 067 is not prime.

One of the dreams of number theorists is to find a simple formula, which for all natural numbers produces only prime numbers. One can easily construct polynomial functions which will give as many primes as required, thus

$$f(n) = n^2 - n + 41$$

yields prime numbers for all naturals $n < 41$. However,

$$f(41) = (41)^2$$

Again,

$$F(n) = n^2 - 79n + 1601$$

yields primes for all $n < 80$.

In about 1640 Pierre de Fermat suspected that

$$P(n) = 2^{2^n} + 1$$

always yielded primes for all natural numbers n. Unfortunately, as was subsequently discovered, this is not so

$$\begin{aligned} P(5) &= 2^{2^5} + 1 = 2^{32} + 1 \\ &= 4\,294\,967\,297 \end{aligned}$$

was factorized by the Swiss mathematician, Leonhard Euler, into

$$= 641 \times 6\,700\,417$$

A proof similar to Euclid's will, however, show that certain sequences contain infinitely many primes. Since every prime after 2 is odd, consider the sequence

$$3 \quad 7 \quad 11 \quad 15 \quad 19 \quad 23 \quad 27 \ldots$$

This sequence increases (term by term) by 4. Hence we have,

3	3 + 4	3 + 2.4	3 + 3.4 ...	
1st	2nd	3rd	4th	...

Clearly, the nth term of this sequence is

$$\begin{aligned} &3 + (n-1)4 \\ &= 3 + 4n - 4 \\ &= 4n - 1 \end{aligned}$$

We assume first of all that this sequence contains some primes, and that the largest such prime is p_n. Hence, the primes contained in this sequence are

$$3, 7, 11, 19, 23, \ldots, p_n$$

Consider the number N where,

$$N = 4(3.7.11.19.23 \ldots p_n) - 1$$

This number is of the form $4n - 1$ and so it is itself a member of the sequence. However, N is not divisible by any of the primes

$$3, 7, 11 \dots p_n$$

in the sequence. Hence, it is either a prime, or it is divisible by a prime *not* of the form $4n - 1$. Now any natural number must be of one of the forms

$$4n - 1, 4n, 4n + 1, 4n + 2$$

since the first number leaves a remainder of 3 when divided by 4, the second is divisible by 4, the third leaves a remainder of 1, and the last leaves a remainder of 2 when divided by 4. Since no other remainder is possible upon division by 4, every natural number must be of one of these forms. Now

$$4n \text{ and } 4n + 2$$

are even, and so cannot be primes. Hence the only primes (other than 2) are of the form

$$4n - 1 \text{ and } 4n + 1$$

Hence, if N is not a prime, it must be divisible by a prime of the form $4n + 1$. However, not every prime divisor of N can be of this form, since

$$(4n + 1)(4m + 1) = 4(4nm + n + m) + 1$$

indicating that the product of two such primes is a number of the same form. Hence, N must have a prime divisor of the form $4n - 1$ which is different from

$$3, 7, 11, \dots p_n$$

This proves that p_n is not the largest prime is the sequence, and so there are infinitely many primes in the sequence.

Q 5.12 Prove, by a similar argument, that there are infinitely many primes in the sequence,

$$5 \quad 11 \quad 17 \quad 23 \quad 29 \dots$$

Q 5.13 Prove that one can always find n consecutive non-prime (*composi-tive*) integers, however great n may be.

Further investigations

1 Let a be the number defined by,

$$a = \sqrt{2 + \sqrt{2 + \sqrt{2 + \dots}}}$$

Prove that

$$a^4 - 4a^2 - a + 2 = 0$$

(i) Explain why $a^2 - a - 2$ must be a factor of this equation.
(ii) Find the other factor, and hence find two more values of a.

2 (i) Show that if x is the side of a regular polygon inscribed in a circle of unit radius, then

$$y = \sqrt{\{2 - \sqrt{(4 - x^2)}\}}$$

is the side of a regular inscribed polygon, having twice the number of sides.
(ii) Use this formula to find an approximation for π.

3 Evaluate the number b, defined by

$$b = \cfrac{1}{1 + \cfrac{1}{1 + \cfrac{1}{1 + \cfrac{1}{1 + \ldots}}}}$$

4 (i) Show that if p/q is an approximation to $\sqrt{2}$, then

$$\frac{p^2 + 2q^2}{2pq}$$

is a better one.
(ii) By starting with $p = q = 1$, show how to construct a sequence of rational approximations to $\sqrt{2}$, and find an expression for the likely error of each of these approximations.

5 The function $f: N \times N \longrightarrow N$ maps an ordered pair of natural numbers onto a single natural number, by the rule

$$f(a,b) = 2^{a-1}(2b - 1)$$

(i) Show that f is a one–one mapping.
(ii) What is the cardinal number of the set,

$$N \times N \times N \times N \times N \times \ldots \times N$$

(n terms)

6 The Mersenne numbers, are numbers of the form

$$M_n = 2^n - 1$$

(i) Show that if n is not prime, then M_n is not prime.
(ii) Is it true that n prime means that M_n prime? (*hint*: consider M_{11}).

7 Perhaps the most amazing result yet found concerning the distribution of primes is the so-called *prime number theorem*. Suppose that A_n denotes the number of primes less than n, then

$$\frac{A_n \log_e n}{n}$$

approaches 1 as n becomes larger and larger. Suppose that we take

$$A_n = \frac{n}{\log_e n} \text{ as an approximation}$$

Use this formula to find the number of primes with ten digits.

8 Another way of determining whether a given number is prime or not is shown below. Let N be the given number, and suppose that n is the least number for which

$$n^2 > N$$

Form the numbers

$$n^2 - N, (n+1)^2 - N, (n+2)^2 - N$$

until one of these is reached which is a perfect square. Let the perfect square reached be x. Then

$$(n+r)^2 - N = x^2$$
$$N = (n+r+x)(n+r-x)$$

(i) Determine whether 8453 is a prime number.
(ii) What happens when N is a prime number?

9 Prove that every odd prime can be expressed as the difference of two squares in one and only one way.

10 Let a and b be natural numbers and d their *greatest common divisor*. We shall write,

$$d = (a, b)$$

Now suppose that we divide b into a giving a quotient q_1 and remainder r_1, then

$$a = q_1 b + r_1 \quad \text{where} \quad r_1 < b$$

We can now divide b by r_1 giving

$$b = q_2 r_1 + r_2 \quad \text{where} \quad r_2 < r_1$$

and continuing in this way we obtain

$$r_1 = q_3 r_2 + r_3 \quad \text{where} \quad r_3 < r_2$$
$$\ldots$$
$$r_{n-2} = q_n r_{n-1} + r_n \quad \text{where} \quad r_n < r_{n-1}$$

and $$r_{n-1} = q_{n+1} r_n + 0$$

This last remainder must be zero at some point because the sequence $r_1, r_2, r_3 \ldots$ is a decreasing sequence of positive natural numbers. Prove that
(i) $d = r_n$ and
(ii) d may be written in the form,
$d = (\text{integer})\, a + (\text{integer})b$
(This process of repeated division is called *Euclid's division algorithm.*)

11 Using the second part of the previous question prove that: if p is prime and divides the product ab, where a and b are two positive integers, then either p divides a, or p divides b, or both.

12 Why is 1 not regarded as a prime number?

6 Modular arithmetic

6.1 Congruences

Another application of the winding function is very useful in gaining insights into the number system. Consider the number line shown in Figure 6.1.

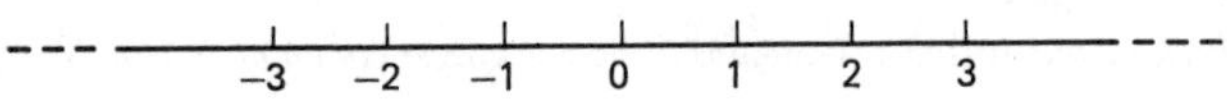

Fig. 6.1 Number line for … − 3, − 2, − 1, 0, 1, 2, 3, etc.

If we map this number line onto a circle of circumference 2 we obtain the diagrams shown in Figure 6.2.

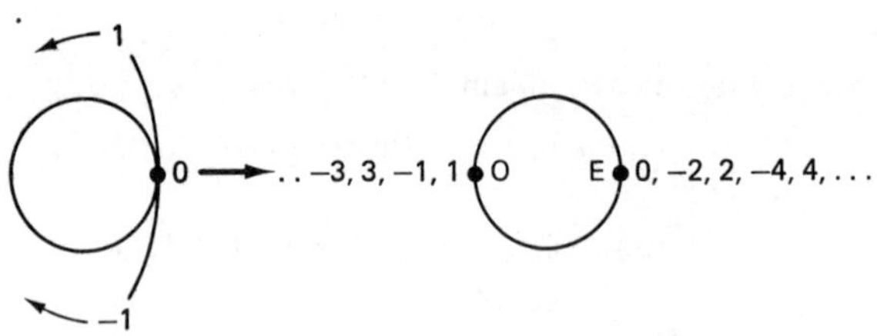

Fig. 6.2 Mapping Figure 6.1 onto a circle of circumference 2

Can you see that each of the integers, … − 3, − 2, − 1, 0, 1, 2, 3, … is mapped onto one of the points labelled E or O? We shall say that any two numbers which are mapped onto the same point under such a mapping are *congruent*, and write

$$a \equiv b \pmod 2$$

2 being the circumference (modulus) of the circle.

Clearly, this mapping splits up the integers into two groups; those which are *even* and those which are *odd*.

Let us map the same number line onto a circle of circumference m (Figure 6.3). The numbers, $1, 2, 3, \ldots, m$ determine m points on the circumference of the circle, but again every integer is mapped onto one of these points.

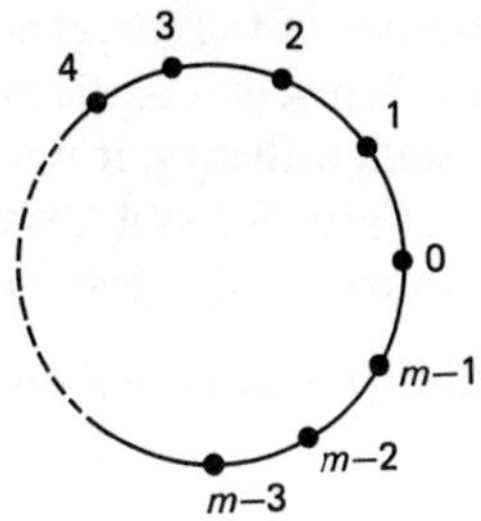

Fig. 6.3 Mapping Figure 6.1 onto a circle of circumference m

For example we may write,

$$m \equiv 0 \pmod{m},\ 2m + 1 \equiv 1 \pmod{m}, \text{ and } m - 1 \equiv -1 \pmod{m},$$

since the number m is mapped onto the same point as 0, the number $2m + 1$ is mapped onto the same point as 1, and $m - 1$ is mapped onto the same point as -1.

If two numbers a and b are mapped onto the same point under this mapping, the numbers must clearly differ by a whole number of circumferences. Hence we have the following:

$$\begin{aligned} \text{if} \quad & a \equiv b \pmod{m} \\ \Rightarrow \quad (a - b) & = pm \text{ where } p \text{ is some integer,} \\ \text{and so,} \quad & a = b + pm \end{aligned} \qquad (1)$$

Notice that this mapping splits up the integers into m groups: that group consisting of the integers divisible by $m, \ldots -2m, -m, 0, m, \ldots$; that group consisting of the integers which leave a remainder of 1 when divided by $m, \ldots, 1 - 2m, 1 - m, 1, m + 1, 2m + 1, \ldots$; the group consisting of the integers which leave a remainder of 2 when divided by $m, \ldots 2 - 2m, 2 - m, 2, m + 2, 2m + 2, \ldots$ etc.

Essentially, then, in dealing with integers congruent to m, we have only m distinct 'numbers', represented by the points on the circle. We can, however, represent these distinct 'numbers' by several different sets. Thus each of the following sets is a suitable set of representatives:

$$\{0, 1, 2, 3, \ldots, m - 1\}$$
$$\{-1, 0, 1, 2, \ldots, m - 2\}$$
$$\{-2, -1, 0, 1, \ldots, m - 3\}$$
$$\ldots$$
$$\{1, 2, 3, 4, \ldots, m\}$$

We call such a set, a *complete set of residues* for the modulus m. In a given situation, one chooses the most suitable set.

By looking at the properties of these residues, we are very often able to deduce corresponding properties of the integers. The connection between the residues and the integers themselves is effected by equation (1) derived above. In order to develop such a theory, it would be extremely useful to be able to deal with a given congruence as if it was an equation. As we shall see, for most purposes we can indeed do this. Suppose that

$$a \equiv b \pmod{m} \text{ and } c \equiv d \pmod{m}$$

then

(a) $$a + c \equiv b + d \pmod{m}$$

(b) $$a - c \equiv b - d \pmod{m}$$

(c) $$ac \equiv bd \pmod{m}$$

To establish these results, we note that if $a \equiv b$ and $c \equiv d \pmod{m}$, then

$$a = b + pm \quad \text{and} \quad c = d + qm$$

for integer p and q.

$$a + c = (b + d) + (p + q)m$$
$$\Rightarrow \quad a + c \equiv b + d \pmod{m}$$

This establishes (a). To establish (c) we have

$$\begin{aligned} ac &= (b + pm)(d + qm) \\ &= bd + bqm + dpm + pqm^2 \\ &= bd + (bq + dp + pqm)m \\ \Rightarrow \quad ac &\equiv bd \pmod{m} \end{aligned}$$

Q 6.1 Establish (b).

We may deduce from this last result (c), that if

$$a \equiv b \pmod{m}$$

then, $$ak \equiv bk \pmod{m}$$

since $k \equiv k \pmod{m}$, or indeed any modulus.

However, we may not divide through a given congruence. For example, consider the numbers 36 and 12. Clearly, we have

$$\begin{aligned} 36 &\equiv 12 \pmod{12} \\ \text{i.e. } 6.6 &\equiv 6.2 \pmod{12} \\ \text{but } \; 6 &\not\equiv 2 \pmod{12} \end{aligned}$$

Q 6.2 Prove that it is permissible to cancel a factor from a congruence only if it and the modulus of the congruence have no common divisors, i.e. they are *relatively prime*.

We shall use congruences to illustrate the proof of a well known divisibility rule. Consider the positive integer N. If N is expressed in the ordinary decimal notation, we can write

$$N = a + b.10 + c.100 + \dots$$

where $a, b, c, \dots$ are the digits of the number N in reverse order.

$$\text{Now} \quad 10 \equiv -1 \pmod{11}$$

and squaring both sides of this congruence, which is permissible by (c), we have:

$$10^2 \equiv (-1)^2 \pmod{11}$$
$$\text{i.e.} \quad 100 \equiv 1 \pmod{11}.$$

Similarly, $1000 \equiv -1 \pmod{11}$ and $10\,000 \equiv 1 \pmod{11}$ etc. Each of these congruences can be multiplied by an integer to give,

$$b.10 \equiv -b \pmod{11}, c.100 \equiv c \pmod{11} \text{ etc.}$$

and then using (A), we have

$$N = a + b.10 + c.100 + \dots \equiv a + (-b) + c + \dots \pmod{11}$$
$$\text{i.e.} \quad N \equiv a - b + c - \dots \pmod{11}$$

Hence N is divisible by 11 only if, $a - b + c - \dots$ is divisible by 11.

Q 6.3 Prove a similar rule for divisibility by 9.

Occasionally, one wants to deal with 'fractions' that occur in congruences. For instance, consider the equation

$$8x \equiv 1 \pmod 3$$

A complete set of residues for the modulus 3 is, 0, 1, and 2. Hence, x can take only these values, or an integer congruent to them, which is the same thing. But,

$$8.0 = 0 \equiv 0 \pmod 3$$
$$8.1 = 8 \equiv 2 \pmod 3$$
$$\text{and} \quad 8.2 = 16 \equiv 1 \pmod 3.$$

Hence, if $8x \equiv 1 \pmod 3$ we conclude that

$$x \equiv 2 \pmod 3$$

We might write this as

$$1/8 \equiv 2 \pmod 3$$

This suggests that the arithmetic associated with the complete set of residues modulus m, behaves like the arithmetic of the integers themselves. However, just as division by 0 is not defined in the integers, or indeed anywhere else, so a similar rule holds for congruences. Consider the equation

$$2x \equiv 1 \pmod{4}$$

A complete set of residues for the modulus 4 is, $-1, 0, 1$, and 2. Now

$$2(-1) = -2 \equiv 2 \pmod{4}$$
$$2(0) = \quad 0 \equiv 0 \pmod{4}$$
$$2(1) = \quad 2 \equiv 2 \pmod{4}$$

and

$$2(2) = \quad 4 \equiv 0 \pmod{4}$$

Hence $2x \equiv 1 \pmod{4}$ is *insoluble*. This means that $\frac{1}{2}$ is not defined in modulus 4 arithmetic. One can easily obtain the following result:

if

$$ax \equiv b \pmod{m}$$

then the congruence is soluble only if a is relatively prime to m, or $b \equiv 0 \pmod{m}$.

Clearly the result is true if $b \equiv 0 \pmod{m}$, for them,

$$ax \equiv 0 \pmod{m}$$

and

$$x \equiv 0 \pmod{m} \text{ satisfies this.}$$

When $b \not\equiv 0 \pmod{m}$, we prove this as follows. Consider a complete set of residues of the modulus m,

$$\{0, 1, 2, 3, \ldots, m-1\}$$

Next consider the set,

$$\{0.a, 1.a, 2.a, 3.a, \ldots, (m-1)a\}$$

i.e.

$$\{0, a, 2a, 3a, \ldots, (m-1)a\}$$

This is also a complete set of residues because each of these numbers is distinct—if they were not, we should have two of them congruent.

i.e.

$$pa \equiv qa \pmod{m}$$

However, since a is relatively prime to m, using the result of Question 1.2 we can cancel it, giving

$$p \equiv q \pmod{m}$$

But, there are only m of these numbers $0, a, 2a, 3a, \ldots, (m-1)a$,

and so

$$p = q$$

Since, then, the set $0, a, 2a, 3a, \ldots (m-1)a$ is a complete set of residues there will be exactly one of them congruent to the given number b.

Q 6.4 In ordinary arithmetic, $3/4 + 2/5 = 23/20$.
(i) What is $3/4 \pmod 3$?
(ii) What is $2/5 \pmod 3$?
(iii) What is $23/20 \pmod 3$?
(iv) Is it true that, $3/4 + 2/5 = 23/20 \pmod 3$?

One immediate application of this theory is to indeterminate equations. Suppose that we wished to solve the equation

$$3x - 4y = 10$$

in integers x and y. We could re-write this,

$$\begin{aligned} & 3x = 10 + 4y \\ \Rightarrow\ & 3x \equiv 10 \pmod 4 \\ \Rightarrow\ & 3x \equiv 2 \pmod 4 \end{aligned}$$

Since 3 is relatively prime to 4, we know that this congruence is soluble. We can now proceed,

$$\begin{aligned} 3.0 &= 0 \equiv 0 \pmod 4 \\ 3.1 &= 3 \equiv 3 \pmod 4 \\ 3.2 &= 6 \equiv 2 \pmod 4 \\ 3.3 &= 9 \equiv 1 \pmod 4 \end{aligned}$$

Hence, if $3x \equiv 2 \pmod 4$ then

$$x \equiv 2 \pmod 4$$

$$\Rightarrow \quad x = 2 + n.4 \text{ where } n \text{ is any integer}$$

Hence,

$$\begin{aligned} 4y &= 3x - 10 \\ &= 3(2 + 4n) - 10 \\ &= 12n - 4 \end{aligned}$$

$$\Rightarrow \quad y = 3n - 1$$

So the solutions in integers of this equation are,

$$\{\ldots(-6, -7), (-2, -4), (2, -1), (6, 2), (10, 5), \ldots\}$$
$$n = -2 \quad n = -1 \quad n = 0 \quad n = 1 \quad n = 2$$

Q 6.5 Prove that the indeterminate equation,

$$2x - 6y = 2n + 1$$

has no integer solutions when $n \neq 0$, and $n \in N$

Another easy application of congruences is to the problem of solving polynomial equations. Consider the cubic equation

$$x^3 - 8x + 6 = 0$$

If we denote by f the function,

$$:x \longrightarrow x^3 - 8x + 6$$

we can examine by congruences, the possible values of the function, $f(x)$. A complete set of residues for the modulus 5 is, 0, 1, 2, 3, and 4.

Then if, $x \equiv 0 \Rightarrow x^2 \equiv 0 \Rightarrow x^3 \equiv 0 \pmod 5$

$$8x \equiv 0 \quad \text{and} \quad 6 \equiv 1 \pmod 5$$

Hence, $f(x) = x^3 - 8x + 6 \equiv 0 - 0 + 1 \pmod 5$

$$\equiv 1 \pmod 5$$

If, $x \equiv 1 \Rightarrow x^2 \equiv 1 \Rightarrow x^3 \equiv 1 \pmod 5$

$$8x \equiv 8 \equiv 3 \quad \text{and} \quad 6 \equiv 1 \pmod 5$$

Hence, $f(x) = x^3 - 8x + 6 \equiv 1 - 3 + 1 \pmod 5$

$$\equiv -1 \equiv 4 \pmod 5$$

Similarly, we can obtain the results,

$x(\equiv)$	0	1	2	3	4
$f(x)(\equiv)$	1	4	3	4	3

(mod 5)

Q 6.6 Show that if $x \equiv 3 \pmod 5$ then $f(x) \equiv 4 \pmod 5$

These results show that

$$f(x) \not\equiv 0 \pmod 5 \text{ for any integer } x.$$

$$\Rightarrow \quad f(x) \neq 0 + 5n \text{ for any integers } x \text{ and } n.$$

Taking the particular value for n of 0, this gives

$$f(x) \neq 0 \text{ for any integer } x$$

Thus we may conclude that the equation

$$x^3 - 8x + 6 = 0$$

has no integer solutions.

In fact, we may conclude considerably more than this. If this equation has no integer solutions it might be supposed that it has rational solutions, i.e. numbers of the form u/v, where u and v are integers. If it does, we may cancel this fraction u/v down to its lowest terms so that u and v are relatively prime. Let us suppose that this has been done. Then

$$(u/v)^3 - 8(u/v) + 6 = 0 \text{ and } u \text{ and } v \text{ are relatively prime.}$$

$$\Rightarrow \quad u^3 - 8uv^2 + 6v^3 = 0$$

$$u^3 = (8uv - 6v^2)v$$

$$\Rightarrow \quad u^3 \equiv 0 \pmod{v}$$

But u and v are relatively prime, and so we may cancel a factor of u^2 out of this congruence, giving

$$u \equiv 0 \pmod{v}$$

$$\Rightarrow \quad u = 0 + nv \text{ for some integer } n$$

$$\Rightarrow \quad u/v = n$$

Hence, if there is a rational solution of this equation, it is in fact an integer. So if there are no integer solutions, there are no rational solutions. We conclude that the equation,

$$x^3 - 8x + 6 = 0$$

has no integer or rational solutions.

The theory presented here is quite general, and the process may be extended to a certain class of polynomial equations. However, only equations whose coefficient of the highest power of x, is 1, can be treated in this way. We call such polynomials *monic*. The general proof is then in two parts, corresponding to the steps above. To begin we let the general monic polynomial equation be

$$x^n + a_{n-1}x^{n-1} + a_{n-2}x^{n-2} + \ldots + a_0 = 0$$

(i) Let $f(x) = x^n + a_{n-1}x^{n-1} + a_{n-2}x^{n-2} + \ldots + a_0$

If $\qquad x \equiv b \pmod{m}$

then, $\qquad x^n \equiv b^n \pmod{m}$ etc

and so $f(x) \equiv b^n + a_{n-1}b^{n-1} + a_{n-2}b^{n-2} + \ldots + a_0 \pmod{m}$

If $\quad f(x) \not\equiv 0 \pmod{m}$ for all $x \in \{0, 1, 2, 3, \ldots, m-1\}$

then

$$f(x) \neq pm \text{ for any integer } p$$

and in particular,

$$f(x) \neq 0 \text{ for any integer } x$$

(ii) Suppose that the equation $f(x) = 0$ has a rational solution u/v reduced to its lowest terms, i.e. u and v relatively prime. Then,

$$\begin{aligned} &(u/v)^n + a_{n-1}(u/v)^{n-1} + a_{n-2}(u/v)^{n-2} + \dots + a_0 = 0 \\ \Rightarrow\quad & u^n + a_{n-1}u^{n-1}v + a_{n-2}u^{n-2}v^2 + \dots + a_0 v^n = 0 \\ \Rightarrow\quad & u^n = -(a_{n-1}u^{n-1} + a_{n-2}u^{n-2}v + \dots + a_0 v^{n-1})v \\ \Rightarrow\quad & u^n \equiv 0 \pmod{v} \end{aligned}$$

Since u and v are relatively prime, we may cancel the factor u^{n-1}, giving

$$\begin{aligned} & u \equiv 0 \pmod{v} \\ \Rightarrow\quad & u = nv \text{ for some integer } n \\ \Rightarrow\quad & u/v = n \text{ an integer} \end{aligned}$$

Q 6.7 Prove that the equation

$$x^n + x^{n-1} + x^{n-2} + \dots + 1 = 0$$

has no rational solutions if n is even

This theory is very useful in examining the roots of monic polynomial equations, but it gives us no clue as to how to deal with non-monic polynomial equations.

Q 6.8 Explain why the general proof fails if the equation is not monic.

There is however, a method of getting round this difficulty. It is best illustrated first by an easy example. Consider the equation,

$$3x^2 - 2x + 2 = 0$$

What we shall do is to transform this non-monic equation into a monic one, by a substitution. Hence we shall write,

$$y = 3x$$

The equation then becomes,

$$3(y/3)^2 - 2(y/3) + 2 = 0$$

i.e.

$$y^2/3 - 2y/3 + 2 = 0$$

and finally

$$y^2 - 2y + 6 = 0$$

Let g be the function

$$g : y \longrightarrow y^2 - 2y + 6$$

A complete set of residues for the modulus 4 is, 0, 1, 2, and 3.

If $y \equiv 0$ we have, $y^2 \equiv 0$, $2y \equiv 0$, and $6 \equiv 2 \pmod{4}$.

Hence, $g(y) = y^2 - 2y + 6 \equiv 0 - 0 + 2$

$$\text{i.e.} \quad g(y) \equiv 2 \pmod 4.$$

Working like this for the complete set of residues, we obtain

$y(\equiv)$	0	1	2	3
$g(y)(\equiv)$	2	1	2	1

Hence, the equation $y^2 - 2y + 6 = 0$ has no rational solutions, because it does not have any integer solutions. This means that,

$$y^2 - 2y + 6 = 0 \Rightarrow y \neq u/v \text{ for any integers } u \text{ and } v$$

and so,

$$9(y/3)^2 - 2(y/3)3 + 6 = 0 \Rightarrow y \neq u/v \text{ for any integers } u \text{ and } v$$

We can now transform back to our original equation, using $y = 3x$. This gives us the following:

$$9(x)^2 - 2(x)3 + 6 = 0 \Rightarrow 3x \neq u/v \text{ for any integers } u \text{ and } v$$
$$\text{i.e.} \quad 9x^2 - 6x + 6 = 0 \Rightarrow x \neq u/3v \text{ for any integers } u \text{ and } v$$
$$3x^2 - 2x + 2 = 0 \Rightarrow x \neq u/w \text{ for any integers } u \text{ and } w$$

Hence our original equation has no rational roots.

The method is easily extended to the general case. Suppose that we were interested in the polynomial equation,

$$a_n x^n + a_{n-1} x^{n-1} + a_{n-2} x^{n-2} + \dots + a_0 = 0$$

We transform this by the substitution

$$y = a_n x$$

giving the new equation

$$a_n (y/a_n)^n + a_{n-1} (y/a_n)^{n-1} + a_{n-2} (y/a_n)^{n-2} + \dots + a_0 = 0$$
$$\text{i.e.} \quad y^n/a_n^{n-1} + a_{n-1} y^{n-1}/a_n^{n-1} + a_{n-2} y^{n-2}/a_n^{n-2} + \dots + a_0 = 0$$
$$y^n + a_{n-1} y^{n-1} + a_{n-2} a_n y^{n-2} + \dots + a_n^{n-1} a_0 = 0$$

This is a monic polynomial, and it is now clear that if this has no integer solutions, it has no rational solutions, and so our original equation cannot have any rational solutions either, because if

$$y \neq u/v \text{ for any integers } u \text{ and } v$$
$$\text{then} \quad a_n x \neq u/v \text{ for any integers } u \text{ and } v$$
$$\Rightarrow \quad x \neq u/w \text{ for any integers } u \text{ and } w$$

Q 6.9 Show that the equation

$$8x^3 + 4x^2 - 4x - 1 = 0$$

has no rational solutions.

6.2 Euclidean constructions

There are a great many construction problems in geometry that can be achieved with compasses and ruler alone. Thus one can construct a square, an equilateral triangle, and bisect an angle. It is always a surprise to learn that no such construction for trisecting an angle is possible. This problem was originally posed by the Greeks, who were much taken to this sort of problem, but for many centuries the question remained unanswered. On the one hand nobody managed to give a construction, but on the other nobody managed to prove that it was impossible. Perhaps the nearest 'solution' that was ever offered is the following. Let us attempt to trisect the angle ABC (Figure 6.4 (a)). The first step is to enclose this angle ABC in a rectangle, and to extend the side DA of this.

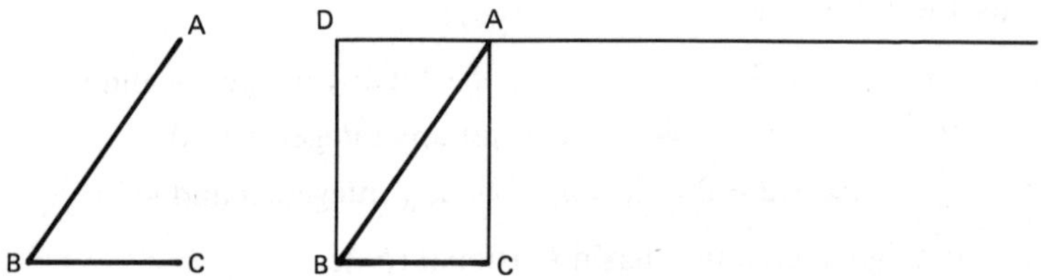

(a)

The second step is to now draw a line through B (Figure 6.4(b)) which cuts AC at E, and DA produced at F so that

$$EF = 2(AB)$$

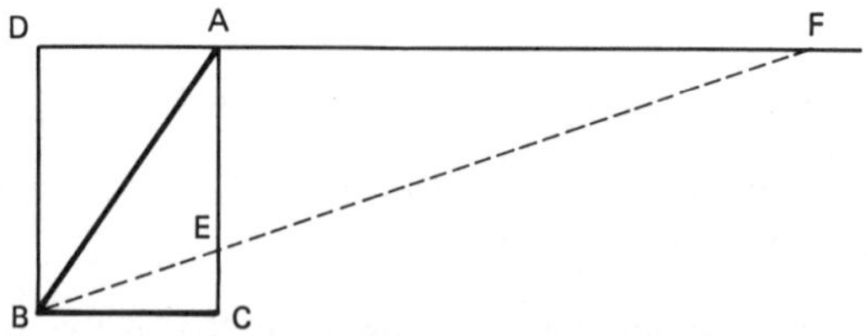

(b)

Fig. 6.4 Attempt to trisect angle ABC

Consider for a moment, any right-angled triangle PQR (Figure 6.5), such that S is the mid-point of the hypotenuse. Since S is the mid-point of PR we may draw a semi-circle on PR as diameter and S as centre. This circle must pass through Q as angle PQR is a right angle (angle in a semi-circle).

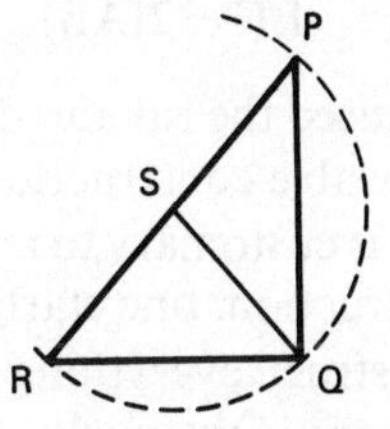

Fig. 6.5 Construct a circle around the right-angled triangle PQR

Hence SQ is a radius of this circle and so

$$SQ = SP = SR$$

Applying this to our diagram for the trisection of angle ABC, let G be the mid-point of EF, and join A to G (Figure 6.6).

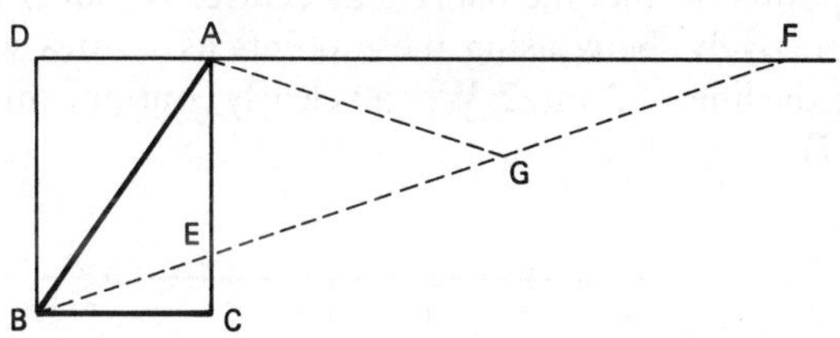

Fig. 6.6 Applying result of Figure 6.5 to right-angled triangle EAF

In the right-angled triangle EAF we can use the result just established to show that

$$EG = GF = AG = AB$$

Hence,

angle ABG = angle AGB (isosceles triangle)

= angle GAF + angle AFG (exterior angle of triangle is the sum of the two interior angles)

But, angle GAF = angle AFG (isosceles triangle)

$\Rightarrow$ angle ABC = 2 angle AFG

But, angle AFG = angle EBC (alternate angles)

$\Rightarrow$ angle ABC = 2 angle EBC

and hence EB trisects angle ABC.

This would seem to dispose of the problem once and for all, but there is, of course, a flaw in this construction. In order to draw the line BF, we used a ruler, and used it in such a way that

$$EF = 2(AB)$$

In other words, we not only used the ruler to draw a straight line, but also to measure. This is not an allowable construction—the ruler can be used only to draw straight lines, and it is customary to refer to such a ruler as a straight edge. In an allowable construction, one starts with a unit length, and uses this, and only this, to construct everything else. Such constructions are called *Euclidean constructions.* Obviously, under such restrictions, our construction is disallowed.

In order to build up a theory to deal with such constructions, we shall have to examine exactly what can, and what cannot be done. Let us suppose that we are given a length to begin with. Then we can construct all rational units of this length.

To see how this is possible, we draw a line with our straight edge. We next mark a point 0 on it, and with the given length set on our compasses, we mark two points on the line using 0 as centre. We label these points, – 1 and 1 respectively. Now using these points as centres we mark two other points on the line, – 2 and 2. We can clearly continue this as far as we wish (Figure 6.7).

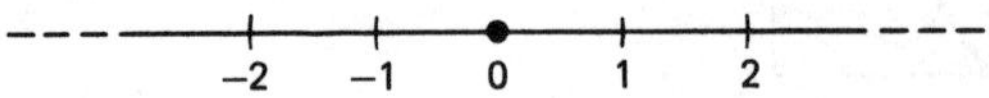

Fig. 6.7 Number line of the integers

Obviously, we have constructed the integers.

If we want the rational number m/n, we locate the integer m on this number line first (Figure 6.8).

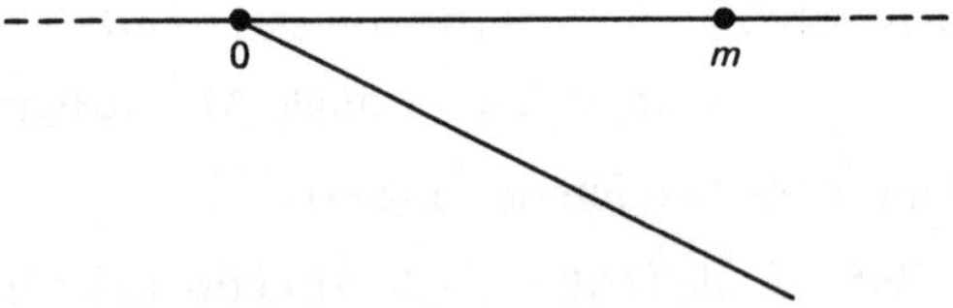

Fig. 6.8 Locate integer m

From 0 we draw any line with our straight edge as shown. We mark on this line the given unit length n times (Figure 6.9).

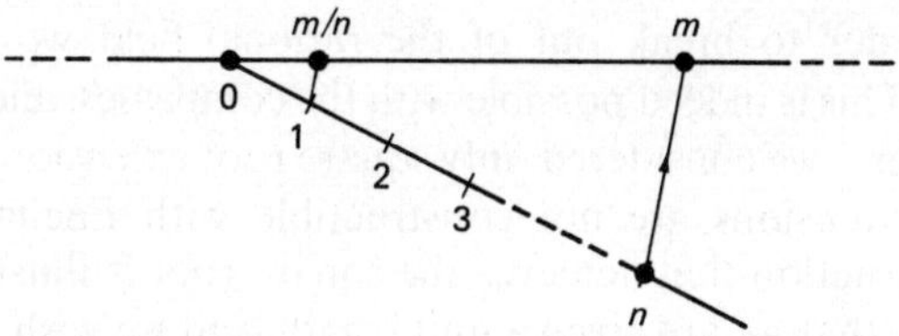

Fig. 6.9 Construct n times unit length on another line

We join the last point on this new line to our point m on the number axis; if we draw a line parallel to this from the point 1 on the new line, the point at which it intersects the number axis represents the number m/n.

Perhaps the only clarification needed is the method whereby the parallel line is constructed. The problem is essentially that shown in Figure 6.10—given a line L and a point p not on the line, construct a line through p, parallel to L.

Choose any points a and b on L. Using a as a centre, set the compasses to a radius of (a, p). Then using b as centre draw an arc on the same side of L as p.

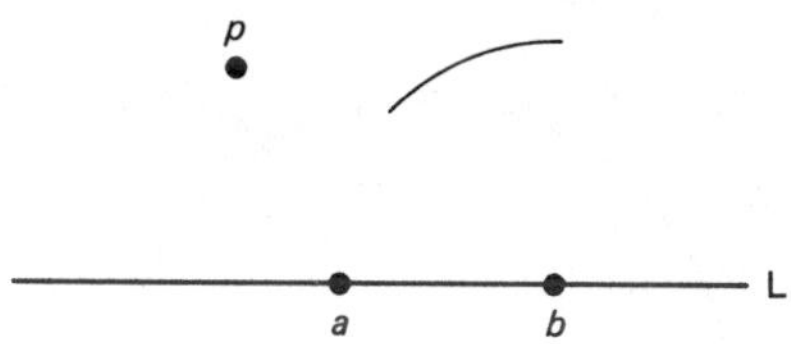

Fig. 6.10 Construction of a line through p parallel to L

Now, using a as a centre (Figure 6.11) set the compasses to radius (a, b). With p as centre, draw an arc to intersect the previous arc at q.

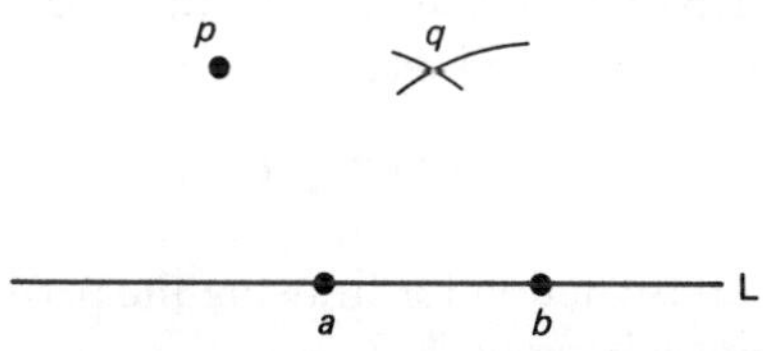

Fig. 6.11 Line drawn through p and q is parallel to L

Joining p to q by the straight edge yields a line through p which is parallel to L.

Since we can construct all numbers of the form m/n from our given unit, we can construct the whole of the rational field R_0.

Q 6.10 Give a construction for ab, where a and b are rational numbers. (You may assume any construction we have already demonstrated.)

Clearly, in order to break out of the rational field we must construct square roots. This is indeed possible with the compasses, and it is the reason that in Chapter 3 we considered only square root extensions of the rational field; other extensions are not constructible with Euclidean tools. The decisive construction that achieves the square root is illustrated in Figure 6.12. Suppose that we are given a unit length, and we wish to construct the length $\sqrt{a}$, where a is any rational number. Obviously, we can construct the length a. We then draw any straight line, and choosing a point p on it, we mark off the length a, giving a point q. Using q as centre, we now mark off the unit length on the line in the same direction, giving a point r.

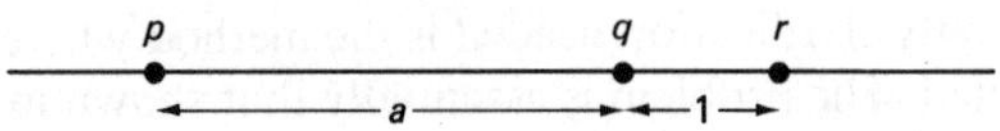

Fig. 6.12 Construction to give the length $\sqrt{a}$

We now bisect the segment (p,r) giving a point O. With O as centre and radius $(O,p) = (O,r)$, we draw a semi-circle (Figure 6.13).

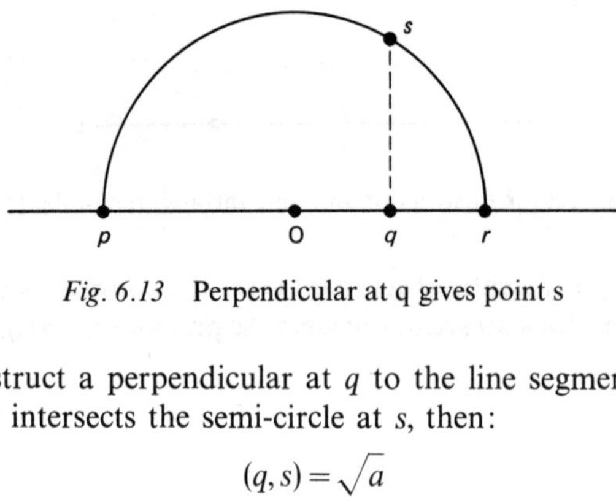

Fig. 6.13 Perpendicular at q gives point s

We now construct a perpendicular at q to the line segment (p,r). If this perpendicular intersects the semi-circle at s, then:

$$(q, s) = \sqrt{a}$$

Q 6.11 (i) Give a construction for drawing the perpendicular to a line L at a point p lying on L,
(ii) Prove that the construction above actually yields $\sqrt{a}$.

Having constructed the square root of any rational number, it is clear that not only can we construct all elements of the rational field R_0, but any element of a square root extension field R_n. The smallest number n for which an element is a member of R_n gives an index of the difficulty of the construction. The theorem in Chapter 3, p. 44 can now be re-interpreted in terms of Euclidean constructions. We have shown that all Euclidean constructible lengths are those that lie in some square root field extension of the rationals. The roots of a cubic equation which are not rational do not lie in such a field, hence they cannot be constructible.

We can now dispose of the trisection problem. In the exercises at the end of Chapter 4, p. 62, the following results were established:

$$\cos 2x = 2\cos^2 x - 1$$
$$\cos 3x = 4\cos^3 x - 3\cos x$$
$$\cos 4x = 8\cos^4 x - 8\cos^2 x + 1$$

Suppose that an angle of 60° can be trisected by a Euclidean construction. Using the second of these equations, this gives

$$\cos 60^\circ = 4\cos^3 20^\circ - 3\cos 20^\circ$$

Now, $\cos 60^\circ = \frac{1}{2}$ and if we let $\cos 20^\circ = x$, we have

$$8x^3 - 6x - 1 = 0 \qquad (1)$$

Let us begin with a unit length. Draw a circle with this length as radius, and any point O as centre (Figure 6.14). Mark any point A on the circumference of this circle, and using A as centre mark another point B on the circle, using as radius, the same unit length. Then angle BOA = 60°. If the angle 60° is trisectable, we can find a point C on this circle so that angle COA = 20°. Draw in the perpendicular from C to OA, giving the point D. Then,

$$OD = \cos 20^\circ = x$$

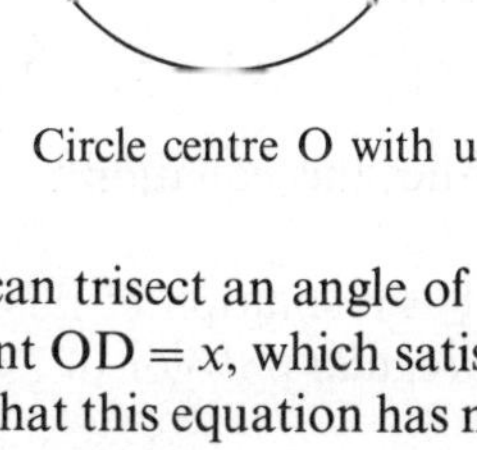

Fig. 6.14 Circle centre O with unit radius

It now follows that if we can trisect an angle of 60° we can construct with Euclidean tools the segment OD = x, which satisfies equation (1).

However, we can show that this equation has no rational roots. Applying now our theorem, we deduce that these roots are not constructible. Hence, an angle of 60° cannot be trisected.

It should be noted that we have *not* proved that no angle can be trisected with Euclidean tools, but only that *not all* angles can be trisected. The truth of the matter is that 90°, and an infinite number of other angles can be so trisected.

The method we have adopted of showing that some angles cannot be trisected is easily adapted to other problems. To show that a given construction is impossible, it suffices to show that some length connected with the construction is a root of a cubic equation with no rational roots.

Q 6.12 Given the fact that it is impossible to trisect an angle of 60°, show that it is impossible to construct a regular polygon with 9 sides.

We may show that it is impossible to construct a regular polygon with 7 sides. Let us suppose for a moment that from a unit length we can construct such a polygon (Figure 6.15); clearly such a polygon can be circumscribed by the unit circle. Then the side AB of such a polygon subtends an angle θ at the centre of this circle, where

$$\begin{aligned} 7\theta &= 360^\circ \\ \Rightarrow \quad 4\theta &= 360 - 3\theta \\ \Rightarrow \quad \cos 4\theta &= \cos(360^\circ - 3\theta) \\ \Rightarrow \quad \cos 4\theta &= \cos 3\theta \\ \Rightarrow \quad 8\cos^4\theta - 8\cos^2\theta + 1 &= 4\cos^3 - 3\cos\theta \end{aligned}$$

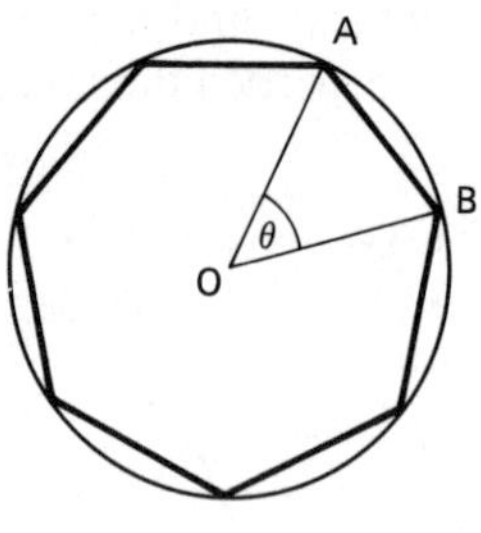

Fig. 6.15 Circle circumscribing 7 sided polygon

If we let $x = \cos\theta$, this reduces to

$$8x^4 - 4x^3 - 8x^2 + 3x + 1 = 0$$

Can you see that $x = 1$ satisfies this equation?

Q 6.13 Explain why $x = 1$ is a solution of this equation, but it is not in any way connected with the regular polygon of 7 sides.

This means that $x - 1$ is a factor of this equation. The other factor is

$$(8x^3 - 4x^2 - 4x - 1)$$

Hence, we need to examine the roots of the equation

$$8x^3 - 4x^2 - 4x - 1 = 0$$

In **Q 6.9** of this chapter, you showed that this equation had no rational roots. It is therefore clear that a regular polygon with 7 sides cannot be constructed with Euclidean tools.

Further investigations

1 Prove that if the regular polygon of m sides can be constructed, then so can the regular polygon of $2m$ sides.
2 Show that a regular polygon of 5 sides can be constructed with Euclidean tools.
3 By considering the constructions for the regular polygons of 3 and 5 sides, deduce a construction for a regular polygon of 15 sides.
4 Given that it is possible to construct a regular polygon with 17 sides, show that it is possible to construct regular polygons with (i) 51, (ii) 85, and (iii) 255 sides.
5 Starting with a unit length, give a Euclidean construction for $\sqrt{(2+\sqrt{2})}$.
6 Is it possible to construct a regular polygon with 24 sides?
7 Given that it is possible to construct regular polygons with p and q sides, under what conditions is it possible to construct a regular polygon of (pq) sides?
8 What is the last digit when the number 3^{89} is evaluated?
9 What is the second to last digit when 3^{89} is evaluated?
10 The Fermat numbers were defined in Chapter 5;

$$P(n) = 2^{2^n} + 1$$

The first few are

5	17	257	65 537	...
$P(1)$	$P(2)$	$P(3)$	$P(4)$	...

This seems to suggest that all Fermat numbers, except the first, end in the digit 7. Prove this.

7 Seeing is not believing

7.1 The development of geometry

The ancient Greek mathematicians were primarily geometers, and the number of geometrical results they amassed is very impressive indeed. Many of the results supposedly attributed to the Greeks were in fact known to people before this time, but this knowledge tended to be isolated and without proof. The major, and lasting, contribution of the Greeks lay in their systematic treatment of this knowledge, and their own discoveries. Much of the credit for this must go to one man, *Euclid*. Euclid compiled a book of geometry called *The Elements*, and as soon as this work appeared it was accorded the highest respect, and from Euclid's successors on to modern times the mere citation of Euclid's book and proposition numbers was regarded as sufficient to identify a particular theorem or construction. No work, except possibly the Bible, has been more widely used, edited, or studied, and probably no work has exercised a greater influence on scientific thinking. Over a thousand editions of Euclid's *Elements* have appeared since the first one printed in 1482, and for more than two millennia, this work has dominated the teaching of geometry.

Euclidean geometry happens to be reasonably descriptive of the everyday world as we understand it (or think we do!). Thus for a long, long time Euclidean geometry was regarded as the only geometry. The essence of Euclidean geometry is not the geometry it contains so much as the method by which this is presented. In fact, Euclid's *Elements* is the prototype of modern mathematical form, by its creation of the postulational, deductive form of thinking. In order to establish a statement in a deductive system one must show that the statement is a necessary logical consequence of some previously established statements. These, in their turn, must be established from some still more previously established statements, and so on. Since the chain cannot be continued backward indefinitely, one must, initially, accept some body of statements without proof. These statements assumed from the start are called *postulates*.

It is not certain what statements Euclid assumed for his postulates nor for that matter exactly how many he had, for changes and additions were made by subsequent editors. However, most editions of *The Elements* distinguished between postulates and axioms. The latter were probably

considered as general truths common to all studies, whereas the former were thought of as pertaining to the special study at hand. In modern mathematics, no such distinction is made, and the words are used interchangeably.
The consensus of historical opinion is that Euclid's probable assumptions were:

Axioms

(1) Things which are equal to the same thing are also equal to each other.
(2) If equals be added to equals, the wholes are equal.
(3) If equals be subtracted from equals, the remainders are equal.
(4) Things which coincide with one another are equal to one another, and said to be *congruent*.
(5) The whole is greater than the part.

Postulates

(1) It is possible to draw a straight line from any point to any other point.
(2) It is possible to produce a finite straight line indefinitely in that straight line.
(3) It is possible to describe a circle with any point as centre and with a radius equal to any finite straight line drawn from the centre.
(4) All right angles are equal to one another.
(5) If a straight line intersects two straight lines so as to make the interior angles on one side of it together less than two right angles, these straight lines will intersect, if indefinitely produced, on the side on which are the angles which are together less than two right angles.

The Elements purports to derive all its 465 theorems from the above ten statements! It would be very remarkable indeed if Euclid's *Elements*, being such an early and colossal attempt at the postulational method of presentation, should be free of logical shortcomings. The most serious defects are the numerous tacit assumptions made by Euclid which do not follow from either the axioms or his postulates. Thus although the first postulate guarantees the existence of at least one straight line joining two points, it does not exclude the possibility of there being other distinct straight lines which also join these points. Euclid frequently assumed that there is a unique straight line joining such points.

Q 7.1 Lewis Carroll, the well known author of the 'Alice' books and a less well known mathematician, parodied these postulates. Carroll's 'First Postulate' is:

(1) Given any two points of view, a controversy may be constructed through them.

Re-write the rest of Euclid's postulates in this form.

Perhaps the greatest tribute that can be paid to Euclid is to remark that no comparable work of mathematics has been attempted until the beginning of this century when Bertrand Russell and Alfred Whitehead wrote *Principia Mathematica*. Their mammoth task was to derive the basis from which all mathematics stems, and to show that this basis was logic itself. In order to understand later developments in geometry, we shall have to pursue some of the work of Euclid. The following is a development very much on the lines of *The Elements*, though much that is not of direct use to us has been excluded.

Theorem 1

If two sides and the included angle of one triangle are equal respectively to the two sides and included angle of another triangle, the triangles are congruent.

Proof

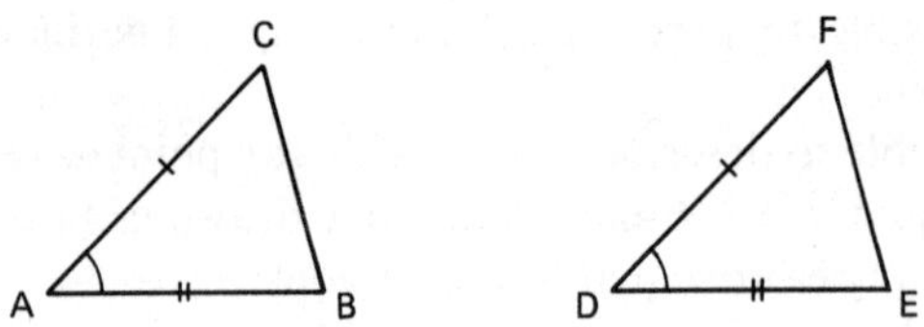

Fig. 7.1 Triangles ABC and DEF with AB = DE, AC = DF, and ∠A = ∠D

Given △ABC and △DEF (Figure 7.1) with AB = DE, AC = DF, and ∠*A* = ∠*D*.

Statements	*Comments*
(1) Place △ABC on △DEF so that point A falls on point D, and AB lies along DE.	(1) There is no assumption of Euclid that such motion of a triangle is possible—in fact it is crucial to his development and characteristic of this geometry.
(2) Point B will fall on point E.	(2) AB = DE, given.
(3) AC will lie along DF.	(3) ∠*A* = ∠*D*, given.
(4) Point C will fall on point F.	(4) AC = DF, given.
(5) BC will coincide with EF.	(5) Postulate 1—Euclid assumed that the line joining two points was unique, although this was nowhere stated by him.
(6) △ABC is congruent to △DEF.	(6) Figures are congruent if they can be made to coincide—axiom 4.

When two triangles ABC and DEF are congruent we shall write,

$$ABC \cong DEF$$

Q 7.2 Prove the following:

Theorem 2

The base angles of an isosceles triangle are equal. (*Hint*: split up the triangle into two other triangles by drawing a line from the vertex to the base which bisects the vertical angle.)

Theorem 3

If two angles and the included side of one triangle are equal respectively to two angles and the included side of another triangle, the triangles are congruent.

Proof

Fig. 7.2 Triangles ABC and DEF with AB = DE, $\angle A = \angle D$, and $\angle B = \angle E$

Given (see Figure 7.2) $\triangle$ABC and $\triangle$DEF with AB = DE, $\angle A = \angle D$, and $\angle B = \angle E$.

Statements	*Comments*
(1) Place $\triangle$ABC on $\triangle$DEF so that point A falls on point D, AB lies along DE.	(1) —
(2) Point B will fall on point E.	(2) AB = DE, given.
(3) AC will lie along DE.	(3) $\angle A = \angle D$, given.
(4) BC will lie along EF.	(4) $\angle B = \angle E$, given.
(5) Point C must fall on point F.	(5) Two straight lines cross in only one point—again Euclid has no assumption concerning this and yet it is clearly needed.
(6) $\triangle ABC \cong \triangle DEF$.	(6) Figures are congruent if they can be made to coincide.

Q 7.3 Prove the following:

Theorem 4

If three sides of one triangle are equal respectively to three sides of another, the triangles are congruent.

Theorem 5

An exterior angle of a triangle is greater than either remote interior angles.

Proof

Given $\triangle$ABC with AB extended to D (postulate 2), forming the *exterior* angle $\angle$CBD of triangle ABC (Figure 7.3)

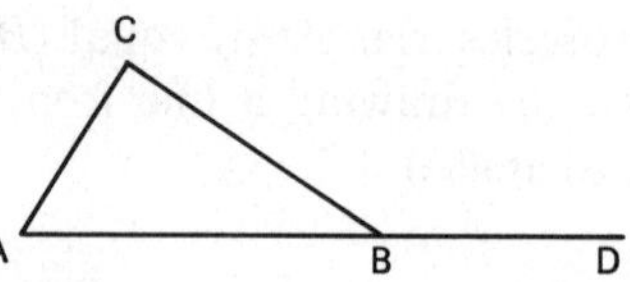

Fig. 7.3 CBD is exterior angle of triangle ABC

Statements	*Comments*
(1) Mark E, the mid-point of CB.	(1) —
(2) Draw AE and extend it to F, making EF = AE.	(2) Postulate 1, 2, and 3.
(3) Draw BF (Figure 7.4).	(3) Postulate 1.

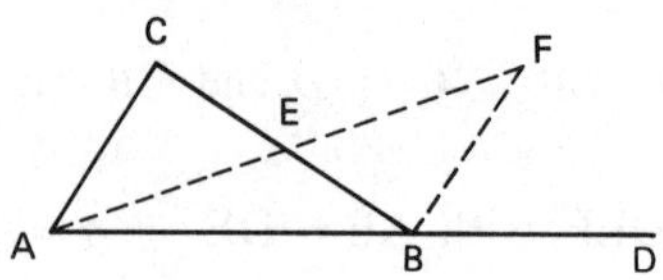

Fig. 7.4

(4) $\angle$CEA = $\angle$FEB.	(4) Rotate EB so that it lies on EF. FEA is a straight line, and BEC is a straight line, and so EC lies on the straight line EA.
(5) $\triangle$AEC $\cong$ $\triangle$BEF.	(5) Theorem 1.
(6) $\angle C = \angle$EBF.	(6) Axiom 4.
(7) $\angle$CBD > $\angle$EBF.	(7) Axiom 5.
(8) $\angle$CBD > $\angle C$.	(8) Axiom 1.

Similarly, using the mid-point of AB and $\angle$ABC, it can be shown that $\angle$CBD > $\angle A$.

Q 7.4 Prove the following:

Theorem 6

Given two lines which are both cut by another line (see Figure 7.5) so that,

$$\angle x = \angle y$$

then the lines are parallel, (i.e. never meet).

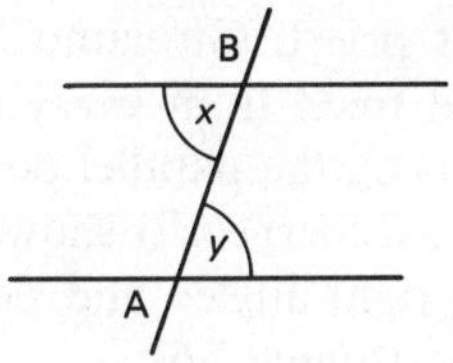

Fig. 7.5 Two lines cut by another, where $\angle x = \angle y$

(*Hint*: assume that the lines meet at a point P. Consider the triangle ABP, and deduce a contradiction.)

So far in our development of geometry we have not made use of postulate 5. There is evidence that a logical development of the theory of parallels gave the early Greeks considerable trouble. Euclid met the difficulties by defining parallel lines as straight lines which do not meet one another however far they may be produced, and by adopting as an assumption his now famous postulate 5. This postulate lacks the terseness of the others and in no sense does it seem 'self-evident'. Actually, it is the converse of Theorem 6, and in this respect it seemed more like a theorem than a postulate. Moreover, Euclid made no use of this postulate until he had proved 29 theorems, indicating that he himself was not happy with it, and reluctant to use it. It was natural to wonder if the postulate was really needed at all, and to think that perhaps it could be derived as a theorem from the remaining nine axioms and postulates. A great deal of effort was spent in investigating this question, and many equivalent forms of the postulate were produced. The most 'self-evident', and the one now commonly quoted in books on Euclid is attributed to John Playfair, a Scottish mathematician. His version states: 'Through a given point can be drawn only one line parallel to a given line.'

Q 7.5 Prove, without using postulate 5, that there is at least one line through a given point, parallel to a given line.

Other possibilities for the postulate are; (1) there exists at least one triangle having the sum of its three angles equal to two right angles, (2) there exists a pair of similar noncongruent triangles, (3) there exists a pair of straight lines everywhere equally distant from one another, and (4) a circle can be passed through any three noncollinear points.

The attempts to derive the parallel postulate as a theorem from the remaining nine axioms and postulates occupied geometers for over two thousand years. Many 'proofs' of the postulate were offered, but each was sooner or later shown to include an assumption equivalent to the postulate itself.

It was not until 1733 that any significant development occurred. In that year, an Italian Jesuit priest, Girolamo Saccheri had published a small book entitled, 'Euclid freed from every flaw'. Saccheri's plan was elegantly simple. Without using the parallel postulate (as postulate 5 has become universally known) Saccheri easily showed that if in a quadrilateral ABCD, angles A and B are right angles, and sides AD and BC are equal, then angles C and D are equal (Figure 7.6).

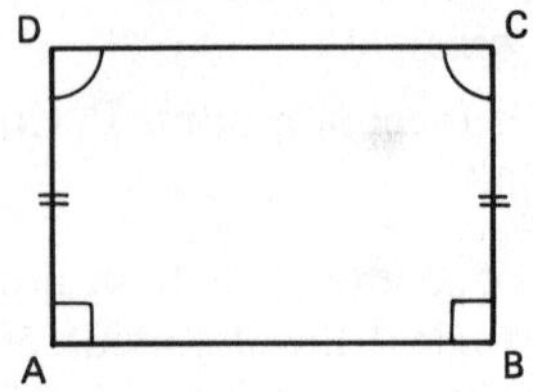

Fig. 7.6 Quadrilateral ABCD

Q 7.6 Prove Saccheri's theorem.

In the Saccheri quadrilateral, then, there are three possibilities: angles C and D are equal
(i) acute angles;
(ii) right angles;
(iii) obtuse angles.
These three possibilities were referred to by Saccheri as the *hypotheses of the acute angle, right angle, and obtuse angle* respectively. He reasoned that if the parallel postulate could be proved from the remaining nine axioms and postulates, then the assumption of the hypothesis of the acute or obtuse angles would lead to a contradiction. Hence, the hypothesis of the right angle must be accepted, and this hypothesis, Saccheri showed, carried with it a proof of the parallel postulate.

Q 7.7 Prove that if the hypothesis of the right angle is accepted, then the parallel postulate can be established.
(*Hint*: consider an equivalent form of the postulate.)

The results that Saccheri managed to obtain were very curious indeed. He proved with little difficulty that the hypothesis of the obtuse angle led to a contradiction. However, try as he might, he could find no contradiction associated with the hypothesis of the acute angle. So convinced was Saccheri that such a contradiction existed that he constructed a rather 'hazy' proof purporting to exhibit one.

Fifty years after the publication of Saccheri's book, the German Johann Lambert, published a book of a similar investigation. Lambert also dealt with a quadrilateral as the basis of his research into the parallel postulate.

The quadrilateral had three right angles, and he considered the hypotheses according as the fourth was acute, right, or obtuse. Thus, with Saccheri, he showed that in adopting these three hypotheses the sum of the angles of a triangle is less than, equal to, or greater than two right angles, respectively. Before some examples of Lambert's work is given, it is as well to review geometry as it was then known. Figure 7.7 illustrates the tentative steps that Saccheri and Lambert made in the discovery of what is now known as non-Euclidean geometry.

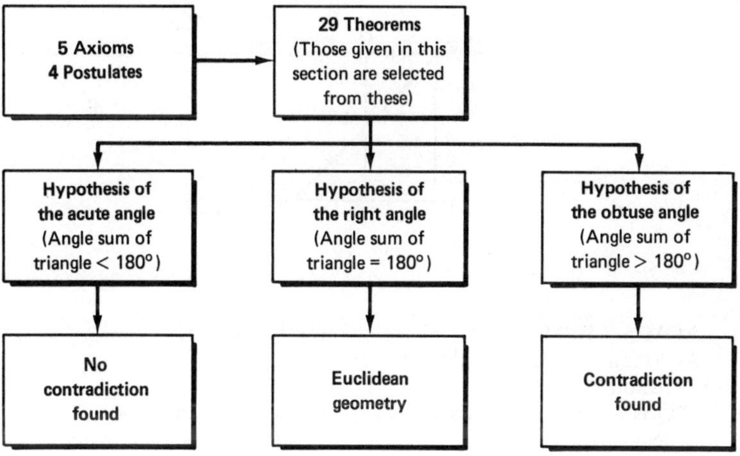

Fig. 7.7 Saccheri and Lambert development of non-Euclidean geometry

Theorem 7

Under the hypothesis of the acute angle, the sum of the angles of any right triangle is less than two right angles.

Proof

Let ABC be any right triangle (Figure 7.8)

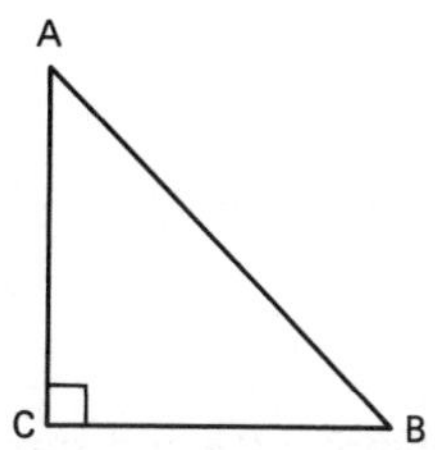

Fig. 7.8 ABC, a right-angled triangle

Statements	*Comments*
(1) Let M be the mid-point of AB.	(1) —
(2) At A construct ∠BAD = ∠ABC.	(2) This is an allowable Euclidean construction. How can it be done?
(3) From M draw MP perpendicular to CB.	(3) It is easy to prove, without the parallel postulate, that only one perpendicular may be drawn to a line from a point.
(4) On AD, mark off AQ = PB.	(4)
(5) Join M to Q (Figure 7.9).	(5)

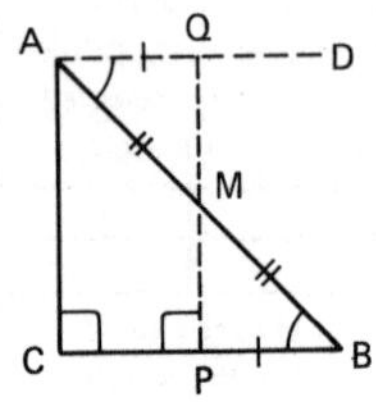

Fig. 7.9

(6) △AQM ≅ △BPM.	(6) Theorem 1.
(7) ∠AQM is a right angle, and points Q, M, and P are collinear.	(7)
(8) ACPQ is a Lambert quadrilateral with acute angle at A.	(8) It has right angles, C, P, and Q. A is acute by the hypothesis of the acute angle.
(9) The angle sum of △ABC = ∠C + (∠CAB + ∠ABC) = 90° + (∠CAB + ∠MAQ).	(9) By construction.
(10) ∠CAB + ∠MAQ = ∠A of the quad. ACPQ.	(10) By construction.
(11) The angle sum of △ABC is less than two right angles.	(11) Hypothesis of the acute angle.

Q 7.8 Prove, using the hypothesis of the right/obtuse angle, that a right triangle has an angle sum of two right angles/greater than two right angles.

Theorem 8

Under the hypothesis of the acute angle, any triangle has an angle sum less than two right angles.

Proof

Let ABC be any triangle (Figure 7.10) and suppose that angle A is at least as big as angle C and B. (Is it always possible to choose angle A of a triangle thus?)

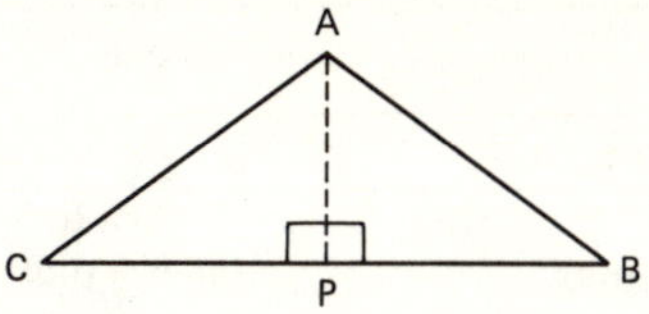

Fig. 7.10 ABC is any triangle. Suppose that $\angle A$ is at least as big as $\angle B$ and $\angle C$

Statements	*Comments*
(1) Construct the perpendicular from A to CB, meeting it at P.	(1) —
(2) Then △APC and △APB are right triangles.	(2) —
(3) The angle sum of △ABC is less than two right angles.	(3) Use Theorem 7.

Q 7.9 Explain why the perpendicular was drawn from the vertex A.

Theorem 9
Under the hypothesis of the acute angle, two triangles are congruent if the three angles of one are equal to the three angles of the other.

Proof
Given two triangles ABC and PQR (Figure 7.11) in which corresponding angles are equal. If AB = PQ, then by Theorem 3, the triangles are congruent. Suppose then that,

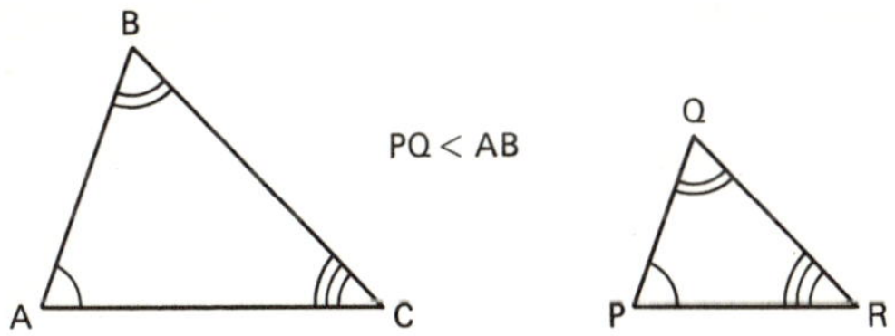

Fig. 7.11 Congruent triangles ABC and PQR, with PQ < AB

Statements	*Comments*
(1) On AB mark off AD = PQ.	(1) —
(2) On AC mark off AE = PR.	(2) —
(3) △ADE ≅ △PQR (Figure 7.12).	(3) Theorem 1 (Figure 7.13).

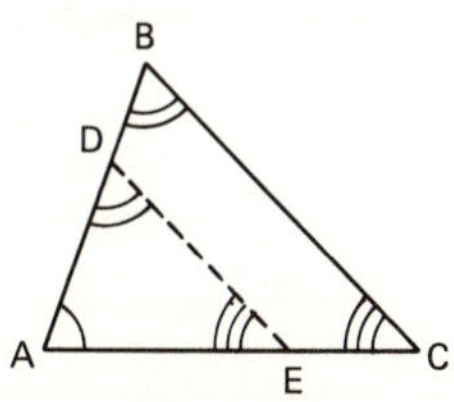

Fig. 7.12

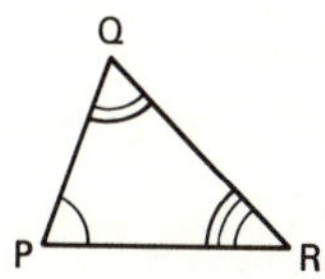

Fig. 7.13

Statements	*Comments*
(4) D lies between A and B.	(4) PQ < AB.
(5) ∠DCA < ∠BCA.	(5) Axiom 5.
(6) E cannot coincide with C.	(6) If it did, ∠DCA = ∠DEA = ∠R = ∠C = ∠BCA which contradicts (5).
(7) E cannot lie on AC produced.	(7) If it did, we should have (Figure 7.14) 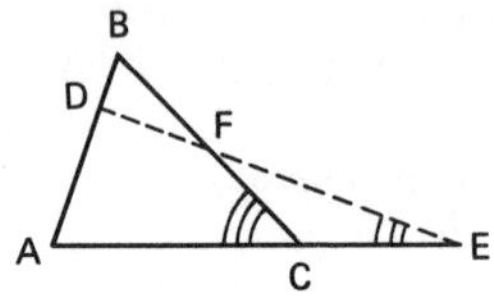 *Fig. 7.14* Then in △FCE, the exterior angle FCA is equal to the remote interior angle FEC, which contradicts Theorem 5 (Figure 7.15).
(8) E lies between A and C.	(8) —
(9) BCED is a quadrilateral.	(9) —
(10) The angle sum of this quad. is four right angles.	(10) 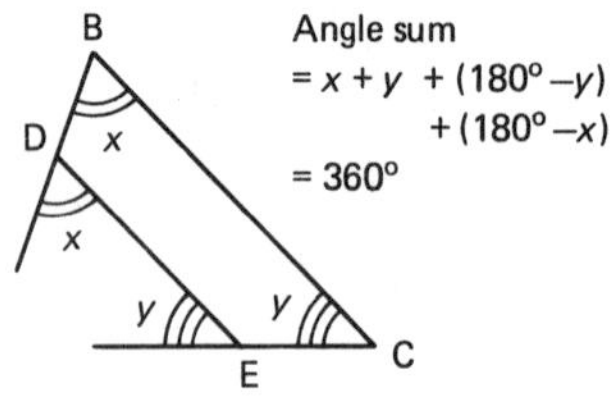 *Fig. 7.15*
(11) By the hypothesis of the acute angle this is impossible.	(11) (i) The angle sum of any triangle is less than 180°, Theorem 8. (ii) Split up quadrilateral into two triangles, and hence its angle sum must be less than 360°.
(12) The original assumption must be false, i.e. PQ ≮ AB.	(12) —
(13) PQ > AB is also false.	(13) Similar proof.
(14) Hence PQ = AB.	(14) —

This theorem goes to show that under the hypothesis of the acute angle, we cannot have figures of the same shape (*similar*) but different sizes.

Q 7.10 Under the hypothesis of the acute angle, we have shown that the angle sum of a triangle is less than two right angles. The amount it differs from this is called its *defect*. Show that if a triangle is split up into two 'sub-triangles' by drawing a line from a vertex to the opposite side, the sum of the defects of these 'sub-triangles' is equal to the defect of the complete triangle.

These theorems seem very strange to us, accustomed as we are to viewing the world through Euclidean spectacles. However, whatever Sachheri and Lambert felt privately, their public conclusions were vague and unsatisfactory, and because of their eagerness to vindicate Euclidean geometry, they cannot be credited with the discovery of non-Euclidean geometry. The credit for this must be shared by three other mathematicians, Gauss of Germany, Johannes Bolyai of Hungry, and the Russian, Nicholai Ivanovitch Lobachevsky. These men approached the subject through the Playfair form of the parallel postulate, by considering the three hypotheses: through a given point can be drawn (i) more than one; (ii) just one; (iii) no lines parallel to a given line. These are, of course, equivalent to the hypotheses of Saccheri and Lambert. Assuming the infinitude of a straight line, the third hypothesis is soon discarded—however, if this is not assumed this hypothesis cannot be discarded.

Q 7.11 In **Q 7.5** you were asked to 'prove' that there is at least one line through a given point parallel to a given line. Look back at your proof and find where the infinitude of the straight line was implicitly assumed.

The remaining two hypotheses both produce sets of theorems, which though strikingly different, are both self-consistent. It is likely that Gauss was the first to suspect that the first hypothesis and the geometry it produced was just as valid as the classical geometry of Euclid, obtained with the second hypothesis. Unfortunately, however, he failed to publish anything on the matter throughout his life, and it was only by an examination of his papers after his death that this research was discovered. Be this as it may, the honour of first publishing these conclusions and a large body of the theorems of non-Euclidean geometry, must be shared by Bolyai and Lobachevsky, who worked independently. Finally it became clear that the parallel postulate was completely independent of the other postulates of Euclid, and that other forms of this postulate would result in different geometries, just as valid, even though they seemed to violate commonsense. This particular story concludes with a discovery made by a student of Gauss, George Bernhard Riemann, that if the infinitude of a straight line is discarded and merely its boundlessness be assumed, then another consistent non-Euclidean geometry can be developed from the third hypothesis.

7.2 Geometry and reality

We have just seen that three perfectly consistent, though radically different, geometries can be developed merely by taking different forms of the parallel postulate of Euclid. Which geometry best describes the world we live in? The answer would seem to be obvious, and should not be too difficult to prove. Each geometry has a proposition concerning the angle sum of a triangle; Lobachevsky and Bolyai's geometry has triangles whose angle sum is less than 180°, Euclidean geometry has triangles whose sum is 180°, while Riemannian geometry has triangles whose sum is greater than 180°. If an experiment is performed to measure the angles of a triangle with very accurate measuring devices, the matter ought to be settled. Experiments already conducted give a sum of 180° within the limits of experimental error. Had the result been noticeably less, or noticeably greater, than this, the consequence would have been that one of the non-Euclidean geometries is preferable to describe the real world. But, as it turned out, nothing was settled by these experiments, because for the triangles considered nothing was known about the deviation from 180°. For instance, in triangles whose sides are a few miles in length, the deviation from 180° might be so small as to be undetectable. This suggests that a question of size must inevitably enter the discussion. In fact we have already had one clue in this direction. Consider Question 7.10. You proved that the defect of a triangle was the sum of the defects of its subdivided triangles. If we continue subdividing each of these triangles, it is clear that the defect of a triangle is a function of its size—the smaller the triangle is, the smaller is the defect. Similarly, one may prove in exactly the same way, that under the hypothesis of the obtuse angle, (i.e. Riemannian geometry) the *excess* of a triangle again depends on its size—the smaller the triangle, the smaller the excess. One is thus forced to the conclusion that Euclidean geometry is a good description of the world we live in considered as a microscopic dot in the Universe, but that the geometry of the Universe itself is possibly non-Euclidean.

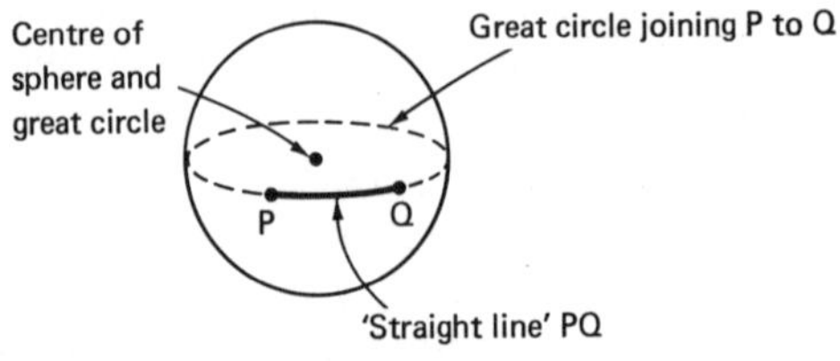

Fig. 7.16 Shortest distance between two points on a sphere is along a great circle

There is a great deal more that can be said on this subject, but it is best illustrated by models. Lambert was the first to notice the resemblance of the geometry following from the hypothesis of the acute angle, to the geometry of a sphere. On the surface of a sphere, 'straight line' must be reinterpreted as the shortest distance between two points, and this is the arc of the great circle joining them (Figure 7.16).

The 'angle' between two 'straight lines' can then also be defined, as Figure 7.17 shows:

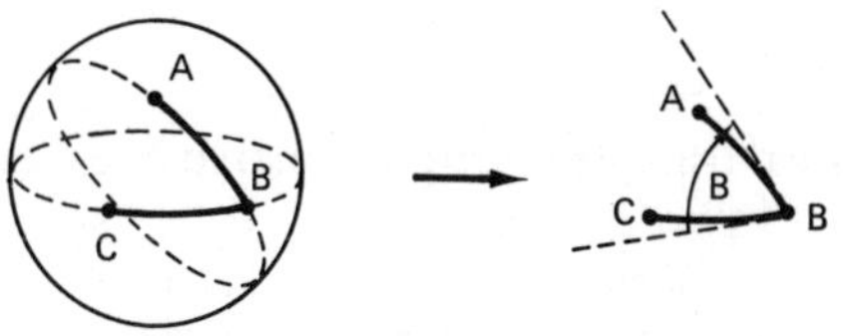

Fig. 7.17 Interpreting the 'angle' between two 'straight lines', on the surface of a sphere

The angle between the 'lines' AB and CB is taken as the angle between the tangents drawn to the great circles AB and CB at B. As usual, we shall letter angles by the letter appearing at its vertex when no confusion then results.

If we digress for a moment and develop some of the theorems of spherical geometry, the correspondence between it and Riemannian geometry becomes clear.

Suppose that the planes of two great circles intersect along a diameter of the sphere (as they must) and cut off on the surface of the sphere two pairs of equal areas called *lunes*. For example, in Figure 7.18, the great circles ABA′B′ and ACA′C′ divide the sphere into four lunes—ABA′CA, ACA′B′A, AB′A′C′A, and AC′A′BA.

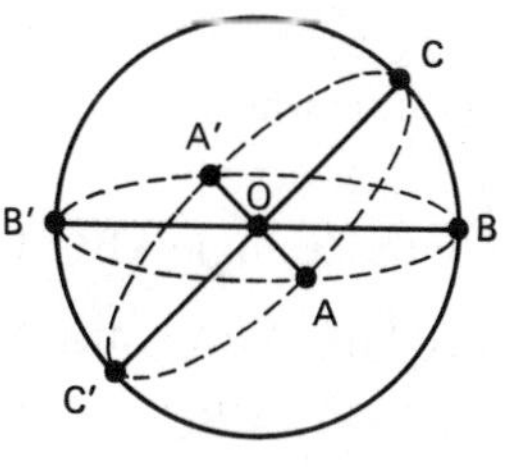

Fig. 7.18 The planes of two great circles

The angle between the planes of the great circles is called the angle of the lune, and clearly, the area of a lune is proportional to this angle. If we denote this angle by θ we have

$$\frac{\text{area of lune}}{\text{surface area of sphere}} = \frac{\theta}{2\pi}$$

$$\Rightarrow \quad \text{area of lune} = 4\pi r^2 \times \frac{\theta}{2\pi}$$

$$= 2r^2\theta \qquad (1)$$

where r is the radius of the sphere.

Theorem 1

The area of a spherical triangle is proportional to its excess.

Proof

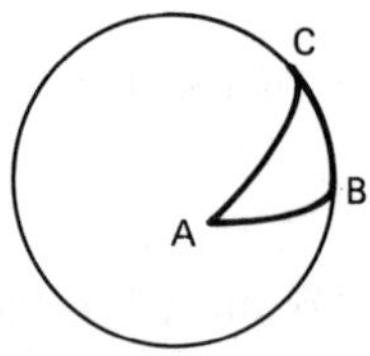

Fig. 7.19 Spherical triangle ABC

Consider any spherical triangle ABC as shown in Figure 7.19. If we draw in the great circles, parts of which form the sides of this triangle, this gives Figure 7.20.

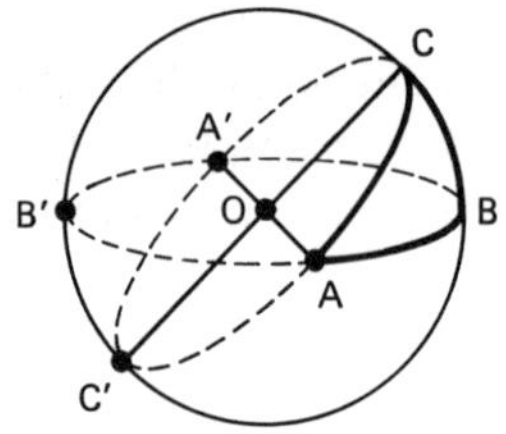

Fig. 7.20 Great circles from which the spherical triangles is made

Let $\triangle$ABC denote the area of the triangle ABC. Using (1) we can write,

$$\text{area of lune ABA'CA} = 2r^2 A$$

$$\Rightarrow \quad \triangle\text{ABC} + \triangle\text{A'BC} = 2r^2 A \qquad (2)$$

$$\text{area of lune BCB'AB} = 2r^2 B$$

$$\Rightarrow \quad \triangle\text{ABC} + \triangle\text{B'CA} = 2r^2 B \qquad (3)$$

$$\text{area of lune CAC'BC} = 2r^2 C$$

$$\Rightarrow \quad \triangle\text{ABC} + \triangle\text{C'AB} = 2r^2 C \qquad (4)$$

If we add (2), (3), and (4) we obtain:

$$2\triangle \text{ABC} + (\triangle \text{ABC} + \triangle \text{A}'\text{BC} + \triangle \text{B}'\text{CA} + \triangle \text{C}'\text{AB}) = 2r^2(A + B + C)$$

However, it is clear from the symmetry of the sphere that

$$\triangle \text{C}'\text{AB} = \triangle \text{CA}'\text{B}', \text{ and so}$$

$$2\triangle \text{ABC} + (\triangle \text{ABC} + \triangle \text{A}'\text{BC} + \triangle \text{B}'\text{CA} + \triangle \text{CA}'\text{B}') = 2r^2(A + B + C)$$

The triangles, ABC, A′BC, B′CA, and CA′B′ make up a hemisphere.

$$\Rightarrow \quad 2\triangle \text{ABC} + 2\pi r^2 = 2r^2(A + B + C)$$

$$\Rightarrow \quad \triangle ABC \qquad = r^2(A + B + C - \pi)$$

But $(A + B + C - \pi)$ is simply the excess of the triangle ABC and so,

$$\triangle \text{ABC} = r^2 E$$

This formula alone suggests the correspondence between spherical geometry and Riemannian geometry because it shows that every spherical triangle must have an angle sum greater than two right angles. (The area of a triangle must be positive.) It also leads to the same conclusion about similar triangles that we proved in the case of Lobachevsky and Bolyai's geometry—that similar triangles of differing sizes are not possible. The proof given there carries over directly into Riemannian geometry, when the third hypothesis is adopted.

Q 7.12 Prove that no two spherical triangles can be similar and not congruent.

Q 7.13 Is it possible for two spherical triangles from different spheres to be similar and not congruent?

The correspondence between the two geometries is thus very strong indeed, and it is an easy matter to demonstrate that it is exact. The point of departure of Riemannian geometry from Euclidean geometry concerns the parallel postulate. If we can show that all the postulates, except the parallel postulate, hold on the surface of a sphere, and that the amended Riemann postulate also holds, then obviously Riemannian geometry and spherical geometry must be identical.

(1) Any two points on a sphere can be joined by a 'straight line' (the great circle joining them).
(2) Clearly it is possible to extend a great circle as far as one chooses (note, however, that the 'straight line' which results while not infinite, is unbounded).
(3) Given any point on a sphere it is possible to describe a circle about this point with the length of any 'straight line' lying on the surface

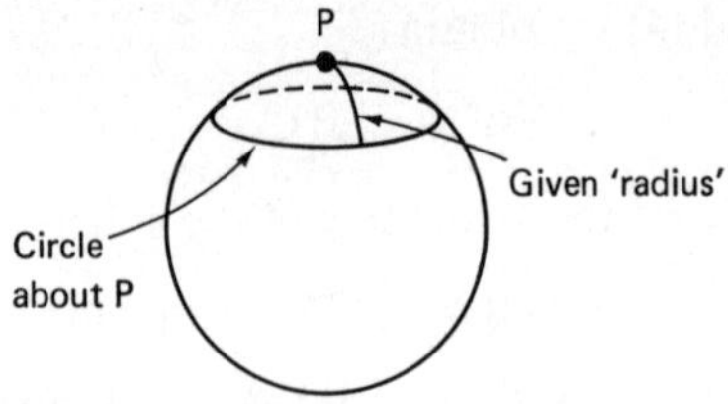

Fig. 7.21 Circle around point P with the length of any 'straight line'

(Figure 7.21). Let P be the given point. Notice that this circle is *not* a great circle, and therefore *not* a 'straight line'.

(4) This is obviously true.

(5) The hypothesis of the obtuse angle is equivalent to: given a point and a line not containing that point, there are no lines through the point parallel (*which do not meet*) to the given line. Let P be the point, and AB the line (Figure 7.22).

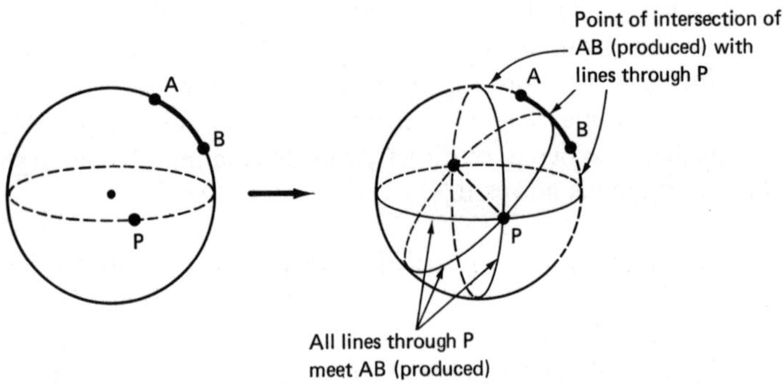

Fig. 7.22 Point P and line AB

We may also interpret the Lobachevsky and Bolyai non-Euclidean geometry by means of a model. This geometry is characterized by the fact that the angle sum of a triangle is less than two right angles. The surface of a *hyperboloid* has such properties. It is clear from Figure 7.23, why this is so.

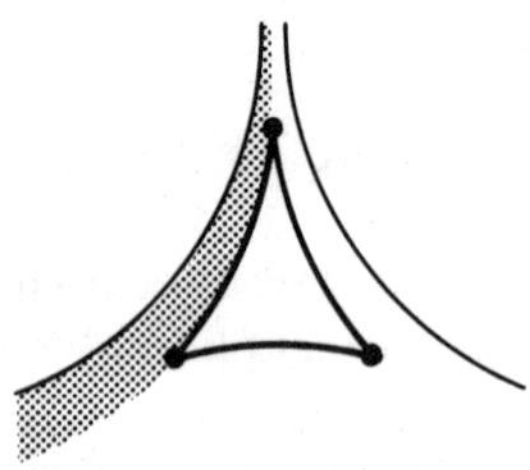

Fig. 7.23 The angle sum of a triangle is less than two right angles

Again, some re-interpretation of straight lines and angles is necessary, and the procedure is to define a 'straight line' as the line of shortest length joining two points, and the 'angle' between two 'straight lines' as the angle between the tangents at the intersection of the 'straight lines'. However, this model is not particularly revealing. Perhaps a better model is the one constructed by Felix Klein. For this model we take a fixed circle S in the Euclidean plane, and interpret the Lobachevsky and Bolyai non-Euclidean plane as the interior of S. A 'point' of this model is a point inside S, and a 'line' is that part of a Euclidean line which is contained in S (see Figure 7.24).

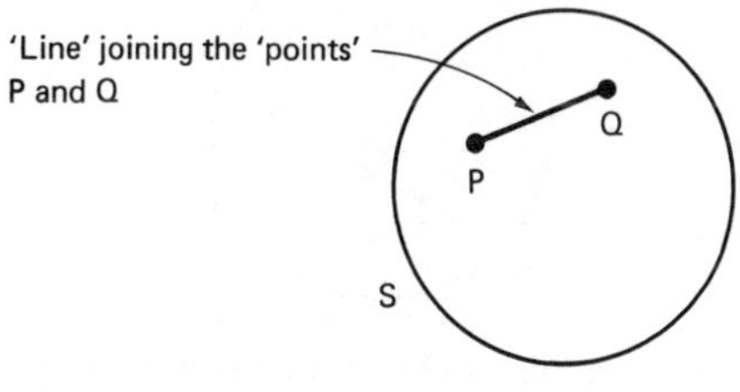

Fig. 7.24 Circle S in the Euclidean plane — the Lobachevsky and Bolyai non-Eulidean plane is the interior of S

It is easy to verify that the first four postulates of Euclid hold. The parallel postulate, of course, does not hold, for now there are many 'straight lines' through a point which do not intersect a given line (Figure 7.25).

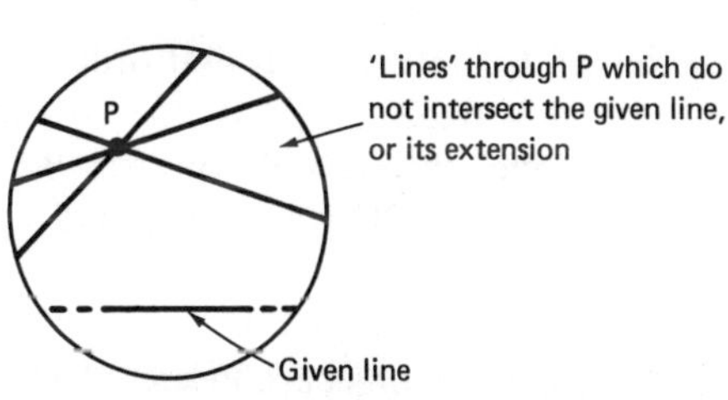

Fig. 7.25 There are many straight lines through a point that do not intersect the given line

Q 7.14 How should 'angle' be interpreted in this model?

Just as one has to be careful when defining 'angle' in this model, so one has to be careful when defining distance. In fact, once distance is defined the problem of defining angle is solved by resorting to trigonometry. Clearly, one cannot define 'distance' in this model as the usual Euclidean distance, for then not only would the angle sum of triangles be 180°, as remarked above, but it would also be possible to construct similar triangles of differing sizes. The problem essentially is that the interior of S is finite and bounded, whereas the Lobachevsky and Bolyai non-Euclidean plane is not. Suppose that P and Q are two 'points' in the model. Continue the segment PQ in both directions (Figure 7.26), till it reaches the boundary of S at the 'points' A and B. We then define:

$$\text{Distance PQ} = \log \frac{(QA)(PB)}{(PA)(QB)}$$

where (QA) etc., are the usual Euclidean distances.

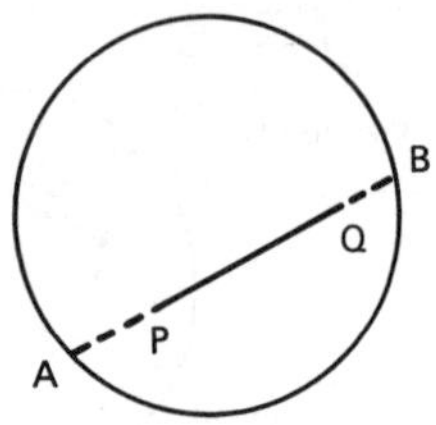

Fig. 7.26 Continuing and line PQ in both directions to the points A and B

Q 7.15 Prove the following properties of this distance function;

(i) distance PQ = distance QP.

(ii) If P, Q, and R lie on a 'straight line' then

$$\text{distance PQ} + \text{distance QR} = \text{distance PR}$$

(Assume that P, Q, and R lie in that order.)

(iii) Let the 'point' P be fixed and let another 'point' Q move along a fixed 'straight line' through P towards B (see Figure 7.25). Show that the distance PQ gets bigger and bigger.

The idea that there may be more than one geometry describing the universe, depending on the size of the objects considered, is a very fruitful one. Riemann generalized geometry as follows. He considered a Universe consisting of a curved surface, which everywhere filled space. This surface was ruled by 'straight lines' which represented the shortest distance joining two points, and called these lines *geodesics.* He imagined that this surface varied from place to place, so that perhaps at one point it was so flat as to be very nearly Euclidean, at others its curvature resembled that of the sphere or hyperboloid and a non-Euclidean geometry described the locale. As fantastic as this seems, there is a growing belief, backed by scientific theories, that this might indeed be the case. In Einstein's General Theory of Relativity, the geometry of space is a Riemann geometry, light travels along geodesics, and the curvature of space is determined by the nature of the matter that fills it.

Appendix

Answers and hints to questions in text

Chapter 1

Q 1.1 Let such a square be

a	*b*
c	*d*

. Then we can write down the following equations.

$$a+b=a+c=b+c=a+d.$$

Hence, $a=c$ and $b=d$ and this does not give a magic square.

Q 1.3 (i) Magic constant $=3(3^2+1)/2=15$.
(ii) This is not possible. Basically there is only one magic square of order 3. However, by rotations and reflections it can be presented in many apparently different forms.

Q 1.4 This is an alternate method of exchanging rows and columns, to give a different magic square.

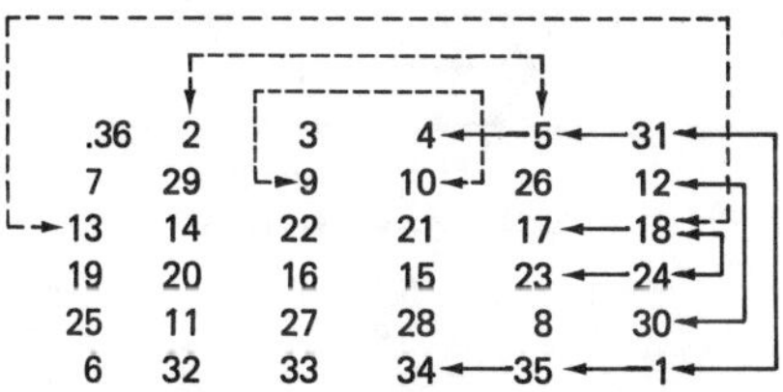

Q 1.6 Suppose the magic square is

a	*b*	*c*
d	*e*	*f*
g	*h*	*i*

. Since the magic constant of the third order magic square is 15, we have:

$$(d+e+f)+(b+e+h)+(a+e+i)+(c+e+g)=4.15$$
$$3e+(a+d+g)+(c+f+i)+(b+h+e)=60$$
$$3e+15+15+15=60 \text{ etc.}$$

Q 1.7 (i) $T_6=6(6+1)/2=21$

Q 1.10 $8T_n + 1 = 8n(n+1)/2 + 1 =$ etc.

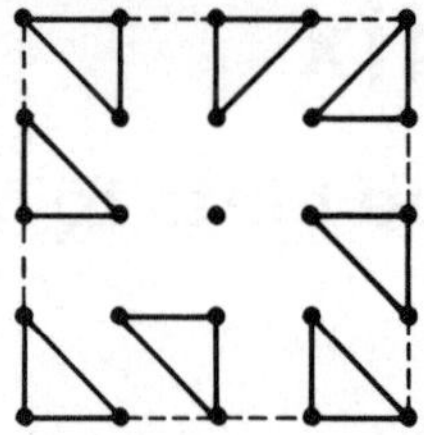

Q 1.11 A decagon has ten sides. Hence, the number of dots is,

$$10_3 = \ldots$$

Q 1.12 (ii) Suppose that the nth tetrahedral number (3_n^{3D}) is $(An^3 + Bn^2 + Cn)/6$ then $6 = A + B + C$; $24 = 8A + 4B + 2C$; and $60 = 27A + 9B + 3C$. Solving these we find that, $A = 1, B = 3$, and $C = 2$. If the nth pyramidal number (4_n^{3D}) is $(An^3 + Bn^2 + Cn)/6$ then $6 = A + B + C$, $30 = 8A + 4B + 2C$, and $84 = 27A + 9B + 3C$. These give, $A = 2, B = 3$, and $C = 1$.

(iii) The generalized formula is

$$M_n^{3D} = ((M-2)n^3 + 3n^2 + (5-M)n)/6$$

Further investigations

1 (i) That a square number is the sum of successive odd numbers.

(ii) $S = n^2$

2 (i)

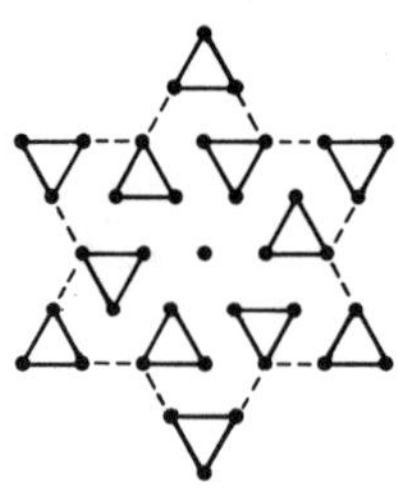

(ii) From this diagram, $S_n = 12T_{n-1} + 1$ (Why T_{n-1} and not T_n?)

$$= \frac{12n(n-1)}{2} + 1 = 6n(n-1) + 1$$

4 Let the total at each vertex be t. The sum of all vertices is then $4t$. Hence,

$$4t = 2(1 + 2 + 3 + 4 + 5 + 6)$$

and so, $t = 21/2$.
Thus it is impossible for a tetrahedron also.

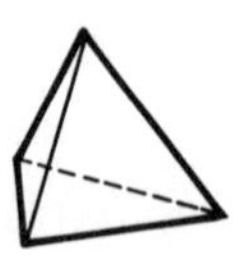

5 The triangular numbers are given by $3_n = (n^2 + n)/2$ and so

$3_n = \{1, 3, 6, 10, 15, 21, 28, 36, \ldots\}$
and $4_n = \{1, 4, 9, 16, 25, 36, \ldots\}$

Thus both 1 and 36 are square and triangular. You might like to check that the next is 1225.

6 $3S_n + 2 = 3(6n(n-1)+1)+2 = 18n^2 - 18n + 5$
$= (3n-1)^2 + (3n-2)^2$
which is the sum of two consecutive squares.

7 $3n^2 + 2 = (n-1)^2 + n^2 + (n+1)^2$
which is the sum of three consecutive squares.

8 We obviously need a number which is both square and a star, multiplied by 3 with 2 added to it. Now,

$n^2 = \{1, 4, 9, 16, 25, 36, 49, 64, 81, 100, \textcircled{121} \ldots\}$
$S_n = \{1, 13, 37, \textcircled{121}, \ldots\}$

Hence, the required number is 121. Then $3 \times 121 + 2$ is the sum of both two and three consecutive squares.

$$\text{i.e.} \quad 365 = 13^2 + 14^2 = 10^2 + 11^2 + 12^2$$

The next square star is 11 881. (Check this). Try to write $3 \times 11\,881 + 2$ in the same way.

Chapter 2

Q 2.1 $179 = 180(n-2)/n \Rightarrow n = 360$
Q 2.2 $m = 6$ and $n = 3$. Hence this tessellation consists of six triangles meeting at a point, and this has already been illustrated.
Q 2.3 $(m-2)(n-2) = 4$. Suppose that $m > 6$, then $m - 2 > 4$. Hence, $n - 2 < 1$, i.e. $n < 3$ and *no* polygon has less than three sides.
Q 2.4

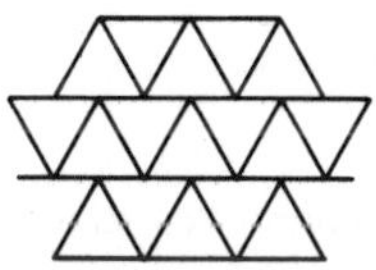

Q 2.5 $(m-1)(n-2) = 2$. Suppose that $m > 3$ then $m - 1 > 2$. Hence, $n - 2 < 1$.
Q 2.6 Because no polygon has two sides, and neither can we have one polygon 'meeting' at a vertex of a polyhedron. Why not $m = 2$?
Q 2.7 $m = 4, n = 3$ represents 4 triangles at each vertex of the polyhedron. It is sketched here. The name given to this polyhedron is *octahedron*.

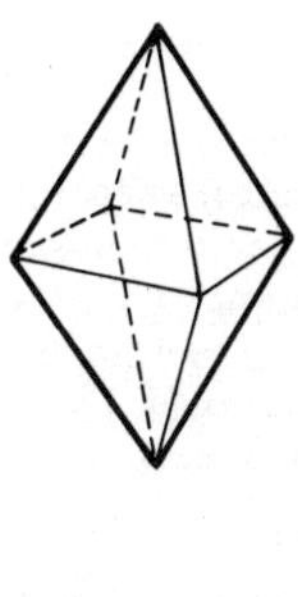

Q 2.8 $(m-2)(n-2)<4$. Suppose that $m>6$, then $m-2>4$. Hence, $n-2<1$ and so $n<3$.
Q 2.9 Because it has six faces.
Q 2.10 In each case you should obtain, $F+V-E=2$. This is called Euler's formula, after the mathematician who first discovered it. It is *not* true for every polyhedron. For instance, the polyhedron shown in the figure (made by sticking pyramids to the faces of a square) does not satisfy this relation,

$$24+14-32 \neq 2$$

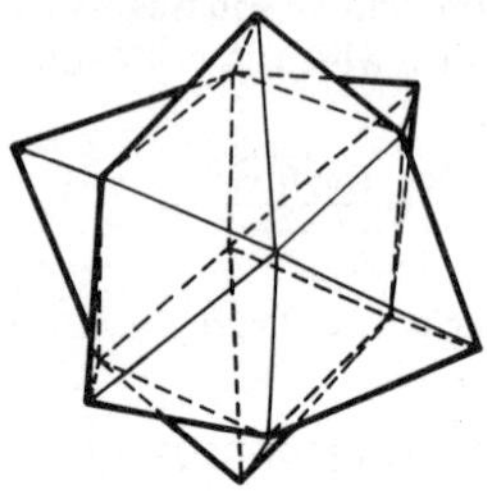

Q 2.11 See **Q 2.6**.
Q 2.12 $V=4/(2-O/180)$. When $m=1, n=3$ and $k=2, h=5$, we have one triangle and two pentagons meeting at a vertex of polyhedron. Hence $O=60^\circ+2\times 108^\circ=276^\circ$. Hence $V=4/(2-276/180)$, which is not a natural number.
Q 2.13 When $m=1, n=3$ and $k=2, h=4$, we have one triangle and two squares meeting at a vertex. Then $O=60^\circ+2.90^\circ=240^\circ$. $V=4/(2-240/180)=6$. This represents a polyhedron. It is in fact a *triangular prism*.
Q 2.14 When $m=1, n=3$, and $k=2$, $h=3$, we have one triangle and two triangles meeting at a vertex, i.e. three triangles. So $O=3.60^\circ=180^\circ$, and then $V=4/(2-180/180)=4$. Hence, this solution represents the tetrahedron.

Chapter 3

Q 3.3 That multiplication distributes over addition means that,

$$ax(b+c)=axb+axc$$

So if addition distributes over multiplication, then

$$a+(b\times c)=(a+b)\times(a+c),$$

which is obviously not true in general.
Q 3.4 (i) $(2,1)\equiv 1$ because $(2,1)$ 'stands' for $2-1$.
(ii) Hence, $(n+1,1)\equiv n$.
Q 3.5 Yes.
Q 3.6 (a,b) corresponds to an integer when b is a divisor of a, or whenever $b=1$. (a,b) corresponds to a natural number whenever it corresponds to an integer ($\neq 0$) and either a and b are both positive or both negative.
Q 3.7 No. For instance the equation $3x=2$ has no solution in the integers (the integers are not closed to division). However, the equation $ax=b$ is always solvable in the rationals if both a and b are rational. This is because the rationals are closed to division, except where the divisor is zero.

Q 3.8 Suppose that $a\sqrt{5}$ is rational, where a is an integer. Then,

$$a\sqrt{5} = p/q \text{ where } p \text{ and } q \text{ are integers. Hence, } \sqrt{5} = p/aq$$

which is equivalent to $\sqrt{5} = p/s$ where p and s are integers, and so $\sqrt{5}$ is rational. Hence, $a\sqrt{5}$ is irrational for all integers a.

Q 3.9 Suppose that $\sqrt{4} = p/q$ where p and q are two integers, and p/q has been reduced to its lowest terms. Then $4 = p^2/q^2$, and so $4q^2 = p^2$. Now the *l.h.s.* is divisible by 4. Hence, the *r.h.s* is also divisible by 4. So p^2 is divisible by 4. *This does not mean that* p is divisible by 4. (Why? e.g. consider 36.)

Q 3.10 Suppose that, $\sqrt{3} = a + b\sqrt{2}$, where a and b are rationals.
Then $3 \quad = a^2 + 2b^2 + 2\sqrt{2}ab$, and so
$\sqrt{2} = (3 - a^2 - 2b^2)/2ab$

The r.h.s of this equation is rational, and hence the l.h.s. must be too. This means that $\sqrt{2}$ is rational. Hence, $\sqrt{3} \notin R_0(\sqrt{2})$.

Q 3.11 Now, $\frac{1}{7} = \dfrac{0.142857\ldots}{7)1.000000\ldots}$ The remainders are respectively, $1, 3, 2, 6, 4, 5, \ldots$

Hence $2/7$ starts with the remainder 2 and so, $2/7 = 0.285714\ldots$ Similarly, $3/7$ starts with the remainder 3 and so gives, $3/7 = 0.428571\ldots$ etc. But we have $2/7 = 2(1/7)$ and $3/7 = 3(1/7)$ etc. This accounts for the property.

Q 3.12 Any element s of $R_{n-1}(\sqrt{c})$ is of the form, $s = x + y\sqrt{c}$ where, $x, y \in R_{n-1}$ and $\sqrt{c} \notin R_{n-1}$. Hence we may write, $s = x + (y/a)a\sqrt{c}$. Since R_{n-1} is a field it is closed to division (other than by zero), and so if $y \in R_{n-1}$ and $a \in R_{n-1}$ then $(y/a) \in R_{n-1}$. Hence, $s = x + z(a\sqrt{c})$ where $x, z \in R_{n-1}$ and $a\sqrt{c} \in R_{n-1}$, and so $s \in R_{n-1}(a\sqrt{c})$. Similarly, any element of $R_{n-1}(a\sqrt{c})$ is also an element of $R_{n-1}(\sqrt{c})$.

Q 3.13 Suppose that $\sqrt[3]{4} = a + b\sqrt{c}$ where $a, b \in R_{n-1}(\sqrt{c})$ and $\sqrt{c} \notin R_{n-1}$, and $R_{n-1}(\sqrt{c})$ is the smallest extension of R which contains $\sqrt[3]{4}$. Then by cubing this expression and re-arranging

$$\sqrt{c} = (4 - a^3 - 3ab^2c)/(3a^2b + b^3c)$$

The l.h.s. of this equation is an element of R_{n-1} since 4, a, b, and c are. By assumption this cannot be; we must have divided by zero giving, $3a^2b + b^3c = 0$. Hence $b = 0$ or $3a^2 + b^2c = 0$. If $b = 0$ then $\sqrt[3]{4} = a \in R_{n-1}$ and so $\sqrt[3]{4}$ is contained within a smaller field than $R_{n-1}(\sqrt{c})$ contrary to assumption. Thus, $3a^2 + b^2c = 0$. Hence, $c = -3a^2/b^2$ and so $\sqrt{c}$ is not real. We conclude that there is no extension containing the number $\sqrt[3]{4}$.

Further investigations

1 The new system is $R_0(\sqrt{3})$, i.e. the rationals extended by the addition of $\sqrt{3}$ and all its rational multiples. This is because:
If $(a, b) = a + b\sqrt{3}$ and $(c, d) = c + d\sqrt{3}$ then,
$(a, b)(c, d) = ac + ad\sqrt{3} + cb\sqrt{3} + 3bd = ac + 3bd + (ad + bc)\sqrt{3}$
$= (ac + 3bd,\ ad + bc)$
The equation $x^2 - 3 = 0$ may be re-written $(a, b)^2 - (3, 0) = (0, 0)$ where a and b are rational. Using the given rules, this means

$$(a, b)(a, b) - (3, 0) = (0, 0) \qquad (a^2 + 3b^2, ab + ba) - (3, 0) = (0, 0)$$

i.e. $(a^2 + 3b^2 - 3, 2ab) = (0, 0)$ and so, $a^2 + 3b^2 - 3 = 0$ and $2ab = 0$.

This second equation yields $a = 0$ or $b = 0$. Suppose that $b = 0$. Then the first equation gives $a^2 - 3 = 0$, which is impossible because a is rational. Hence, $a = 0$.

Then the solution is; $3b^2 - 3 = 0$ and so, $b = \pm 1$ (b is rational as required) and,

$$x = (0, \pm 1)$$

2 Consider the two numbers,

$$x = 0.101\,001\,000\,1\ldots \text{ and } y = 0.010\,110\,111\,0\ldots$$

Obviously x and y are irrational. (Why?) But

$$x + y = 0.111\,111\,11\ldots \text{ which is rational}$$

3 We prove this in two stages.

(i) Since a is an integer, $a^2 - 1$ is an integer. Suppose that, $\sqrt{(a^2 - 1)} = p/q$, a rational number reduced to its lowest terms.

Then, $(a^2 - 1)q^2 = p^2$. Since the r.h.s. is divisible by p, the l.h.s. must be too. However, q cannot be since p/q was reduced to its lowest terms. So p and q contain no common factors. Hence p and q^2 contain no common factors, and so q^2 is not divisible by p. So $(a^2 - 1)$ must be divisible by p. Hence

$$(a^2 - 1) = rp \text{ for some integer } r$$

But, $(a^2 - 1)q^2 = p^2$ and so, $rpq^2 = p^2 \Rightarrow rq = p/q = \sqrt{(a^2 - 1)}$ and hence, $\sqrt{(a^2 - 1)}$ is an integer (rq).

(ii) We now show that it cannot be an integer. Suppose that,

$$\sqrt{(a^2 - 1)} = n, \text{ then } a^2 - 1 = n^2; (a - n)(a + n) = 1.$$

$a - n$ and $a + n$ are integers whose product is 1.
Hence, $a - n = a + n = \pm 1$ and so, $a = \pm 1$ and $n = 0$. But $a > 1$, and we conclude $\sqrt{(a^2 - 1)}$ must be irrational if a is an integer.

4 The equation $x^2 + 1 = 0$ can be re-written,

$$(a,b)^2 + (1,0) = (0,0).$$

(Check that (1,0) behaves as the '1' in this new system.) Using the given rules;

$$(a^2 - b^2 + 1, 2ab) = (0,0) \text{ and so;}$$
$$a^2 - b^2 + 1 = 0 \text{ and } 2ab = 0.$$

As before, if $b = 0$ then $a^2 + 1 = 0$ which is impossible if a is real. Hence, $a = 0$. So $b = \pm 1$ and $x = (0, \pm 1)$.

Since the equation $x^2 + 1 = 0$ has a solution, $x = \pm\sqrt{-1}$, we conclude that $\sqrt{-1} = (0,1)$.

5 We can use our theorem on cubic equations. If $x = \sqrt[3]{a}$, then $x^3 - a = 0$. This is a cubic equation with rational coefficient. Hence if it doesn't have any rational roots, none of its roots lies in any square root extension field of the rationals.

6 $a = \sqrt{2} + 2\sqrt[3]{4}$ and so $(a - \sqrt{2})^3 = 8.4 = 32$. Hence,
$a^3 - 3a^2\sqrt{2} + 3a.2 - 2\sqrt{2} = 32$, i.e. $a^3 + 6a - 32 = (3a^2 + 2)\sqrt{2}$.
So, $(a^3 + 6a - 32)^2 = (3a^2 + 2)^2.2$ which gives,

$$a^6 - 6a^4 - 64a^3 + 12a^2 - 384a + 1016 = 0.$$

Chapter 4

Q 4.1 By symmetry, $W(-k) = (a, -b)$.
Q 4.2 $W(k + 3\pi/2) = (b, a)$.

Q 4.3 Let $W(\pi/8) = (a, b)$. The arcs AB and BO are equal since,

$$\pi/4 - \pi/8 = \pi/8 - 0.$$

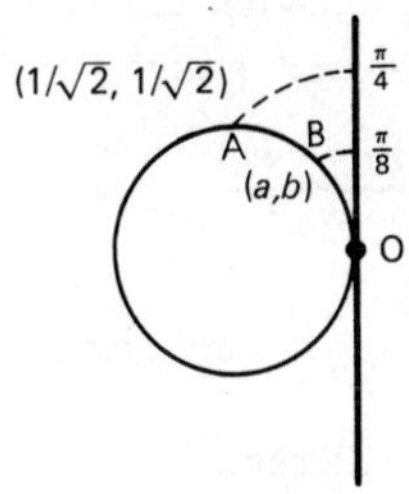

Hence, the corresponding chords are equal and so,

$$\sqrt{\{(1/\sqrt{2} - a)^2 + (1/\sqrt{2} - b)^2\}} = \sqrt{\{(a-1)^2 + (b-0)^2\}}$$
$$\Rightarrow \quad \sqrt{2}a + \sqrt{2}b = 2a$$

But, $(a, b) \in S$ and so $a^2 + b^2 = 1$. Solving these equations simultaneously, we get

$$a = \tfrac{1}{2}\sqrt{(2+\sqrt{2})} \text{ and } b = \tfrac{1}{2}\sqrt{(2-\sqrt{2})}$$

Q 4.4 (i) Let $W(\pi/3) = (x, y)$ and $W(\pi/6) = (a, b)$, Now,

$$\pi/2 - \pi/3 = \pi/6 - 0$$

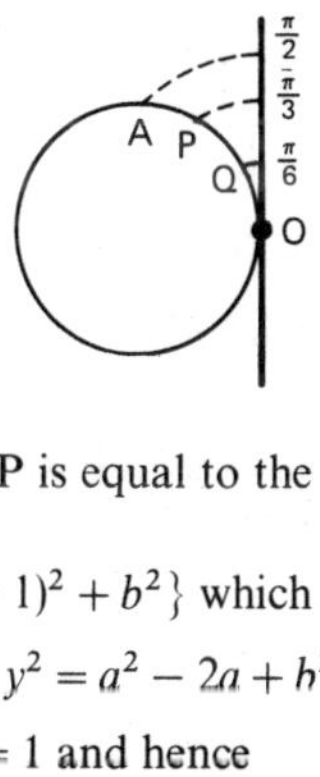

From this last relation, the arc AP is equal to the arc QO. Hence the corresponding chords joining them are too:

$\sqrt{\{(0-x)^2 + (1-y)^2\}} = \sqrt{\{(a-1)^2 + b^2\}}$ which gives,

$$x^2 - 2y + y^2 = a^2 - 2a + b^2$$

Since $P, Q \in S$, $x^2 + y^2 = a^2 + b^2 = 1$ and hence

$$-2y = -2a \text{ i.e. } y = a$$

But, $x^2 + y^2 = 1$ gives $x^2 + a^2 = 1$ $x^2 = 1 - a^2 = b^2$
Hence, $x = b$. (Why not $x = -b$?)
(ii) Obviously, chord PQ = chord QO

$$\sqrt{\{(b-a)^2 + (a-b)^2\}} = \sqrt{\{(a-1)^2 + b^2\}}$$
$$a^2 - 4ab + b^2 = 1 - 2a$$

But, $a^2 + b^2 = 1$ and so,

$$b = \tfrac{1}{2} \text{ and } a = \sqrt{3}/2$$

Hence, $W(\pi/6) = (\sqrt{3}/2, \tfrac{1}{2})$ and $W(\pi/3) = (\tfrac{1}{2}, \sqrt{3}/2)$.

Q 4.5 You should be able to obtain

$$W(k+m) = \frac{ab - pq}{bp - aq}, \frac{p^2 - a^2}{bp - aq}$$

Consider, $\dfrac{ab-pq}{bp-aq}=\dfrac{(ab-pq)(ap-bq)}{(bp-aq)(ap-bq)}$

$$=\frac{(ab-pq)(ap-bq)}{abp^2-pqb^2-pqa^2+abq^2}$$

$$=\frac{(ab-pq)(ap-bq)}{ab(p^2+q^2)-pq(a^2+b^2)}$$

$$=\frac{(ab-pq)(ap-bq)}{ab-pq}=ap-bq$$

The second part requires similar albegra.

Q 4.6 Using the formula just established,

$W(k+m)=(ap-bq, aq+bp)$ if $W(k)=(a,b)$ then $W(2k)=(a^2-b^2, 2ab)$

Q 4.7 Should be easy by now.

Q 4.8 The proof just given is very easily generalized to take care of this. Merely replace $\sqrt{2}$ by x.

Q 4.9 Again it is a simple enough matter to extend the proof to cover this case. Start by considering the set

$$\ldots(34.10^0, 35.10^0), (34.10^1, 35.10^1), \ldots\}$$

Take the logarithms of this set, and map onto a circle of unit circumference. Then consider the set, $\{4, 4^2, 4^3, \ldots\}$.

Further investigations

1 Set 7 ... as a number beginning with 7. Then we require x so that, $7\ldots = 2^x$

$$\Rightarrow \quad \log(7\ldots) = x\log 2$$

Now the decimal part of a logarithm determines the digits of that number. Let us therefore examine the multiples of log 2. However, as you have probably already discovered, it is no good trying for $x < 10$.

Now the decimal part of log (7 ...) is such that

$$\underset{(\log 7)}{0.845\,098\,1} < \text{dec part}(\log(7\ldots)) < \underset{(\log 8)}{.903\,09}$$

So we require a multiple of $\log 2 = 0.301\,03$ which has a decimal part between these two numbers. If you have a calculator with a *constant* key this is easily achieved by continually adding 0.301 03 each time—but remember to keep count! Hence,

$$\begin{aligned} \log 2 &= 0.301\,03 \\ 2\log 2 &= 0.602\,06 \\ 3\log 2 &= 0.903\,09 \\ &\;\;\ldots \\ 56\log 2 &= 16.857\,68 \end{aligned}$$

Hence, the smallest power of 2 beginning with 7 is 2^{56}.

2 Suppose that $\log_a b = p/q$ a rational number. Then $a^{p/q} = b$ and b is a rational power of a.

3 These are simple applications of the formula in this chapter.
4 Now, $100^{19 \cdot 77} > 1977$ because, $100^2 = 10\,000 > 1977$. Hence, $(100^{19 \cdot 77})^{100} > (1977)^{100}$ i.e. $100^{1977} > 1977^{100}$

Chapter 5

Q 5.1 (i) The two values of a occur as a result of the square root sign, which may be either positive or negative.
(ii) 1.990 369 4.

Q 5.2 $b = \frac{1}{2} - \frac{1}{2} + \frac{1}{2} - \frac{1}{2} + \ldots \qquad -b = -\frac{1}{2} + \frac{1}{2} - \frac{1}{2} + \frac{1}{2} - \ldots$
and so, $-b = \frac{1}{2} - \frac{1}{2} + \frac{1}{2} - \frac{1}{2} + \ldots$ (interchanging pairs)
$= \frac{1}{2}$ as in (b). Hence $b = -\frac{1}{2}$.

Q 5.3 1 2 3 4 5 6 ...
1^2 2^2 3^2 4^2 5^2 ...

Q 5.4 Suppose that the set of positive rationals is,

$$R^+ = \{r_1, r_2, r_3, r_4, \ldots\}$$

Then, we may write the complete set of rationals as

$$R = \{0, -r_1, r_1, -r_2, r_2, \ldots\}$$

and they then have the same cardinal number as the naturals.
Q 5.5 Let (a, b) and (c, d) be any two segments of the real numbers. Then geometrically we have Figure A.25, and this clearly shows the two sets in one–one correspondence.

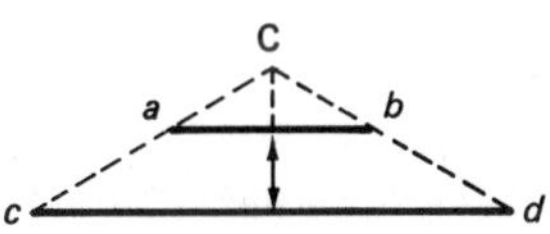

Q 5.6 Suppose A has n elements and that there are $f(n)$ (A function of n) subsets of A. Then the subsets are:

$$\varnothing, \{\ \}, \{\ \}, \{\ \}, \ldots, A$$

Consider another set B formed by adding the single element b to the set A. Then the number of subsets of B is $f(n+1)$, and these subsets are:

$$\varnothing, \{\ \}, \{\ \}, \{\ \}, \ldots, A$$
$$\{b\}, \{\ , b\}, \{\ , b\}, \{\ , b\}, \ldots, B$$

But in each of these lines there are just $f(n)$ sets. Hence,

$$f(n+1) = 2f(n). \text{ But } f(1) = 2 \text{ so } f(2) = 2.2 = 4, f(3) = 8 \ldots$$

and hence, $f(n) = 2^n$ as required.
Q 5.7 Let c be the integer. If c is not prime it has a smallest factor p. Then p is prime, for if p were composite, c would have a still smaller factor.
Q 5.8 The number of divisors of N is $(a_1 + 1)(a_2 + 1) \ldots (a_r + 1)$.
Q 5.9 We test N for prime factors by going through the prime numbers in order. Suppose that the first prime factor of N is p where p is greater than $\sqrt{N}$. Then $N = px$ and x is the other factor of N. But if $p > \sqrt{N}$, then $x < \sqrt{N}$ and so it would be reached before p.

Q 5.11 $2N+1$ is not prime if and only if N occurs in the table. Let $2N+1 = 6\,853\,067$ so that, $N = 3\,426\,533$. Does this number occur in the table? We have to solve the equation

$$(2m+1)n+m = 3\,426\,533 \text{ in natural numbers } n \text{ and } m.$$

Let $n=1$, and then test values of m. Then let $n=2$, and test values of m, etc. For example, with $n=1$,

$$2m+1+m = 3\,426\,533 \qquad 3m = 3\,426\,533$$

which does not give a whole value for m. However, when $n=6$

$$12m+6+m = 3\,426\,533 \Rightarrow 13m = 3\,426\,533, m = 263\,579$$

Hence, 6 853 067 is not prime since 3 426 533 occurs in the 6th row and the 263 579th column of the table.

Q 5.12 We can re-write this sequence,

5	11	17	23	29	...
5	5+6	5+2.6	5+3.6	5+4.6	...

Obviously, the nth term of the sequence is, $5+(n-1)6 = 6n-1$
Note that any number is one of the forms

$$6n-1, 6n, 6n+1, 6n+2, 6n+3, \text{ or } 6n+4.$$

Q 5.13 Consider the sequence of consequitive numbers,

$$(n+1)!+2; (n+1)!+3; \dots; (n+1)!+(n+1)$$

(Here, $(n+1)!$ stands for, $(n+1)(n)(n-1)(n-2)\dots 3.2.1$)

Further investigations

1 $a = \sqrt{(2+\sqrt{\{2+\sqrt{(2+\dots)}\}})} \Rightarrow a^2-2 = \sqrt{(2+\dots)}$
$a^4-4a^2+4 = 2+\sqrt{(2+\dots)}$ i.e. $a^4-4a^2+2 = a$. Hence,

$$a^4-4a^2-a+2=0$$

(i) Clearly, a^2-a-2 is a factor of this because of the working above.
(ii) The other factor is, a^2+a-1, and hence

$$a = (-1 \pm \sqrt{\{1-4.1.(-1)\}})/2 = (-1 \pm \sqrt{5})/2$$

(Proceeding like this, we may find as many different values of a as we choose. In fact there are at least $\aleph_0$ of them. It is easy to verify that if x is one value of this 'number' a, then $\sqrt{(2+x)}$ is another.)

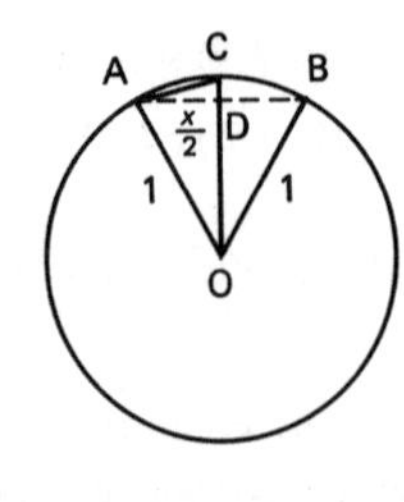

2 (i) Now $OD^2 = 1-(\frac{1}{2}x)^2 = (4-x^2)/4$ and hence $CD = 1-OD$
$= 1-\frac{1}{2}\sqrt{(4-x^2)}$

so, $$y^2 = (\tfrac{1}{2}x)^2 + (1 - \tfrac{1}{2}\sqrt{4 - x^2})^2$$
$$= x^2/4 + 1 - \sqrt{(4 - x^2)} + (4 - x^2)/4$$
$$= 2 - \sqrt{(4 - x^2)}$$

(ii) We shall use the fact that the perimeter of a regular polygon gets closer and closer to the circumference of the circle as the number of sides becomes larger.

Suppose we start with a triangle. Then each side is of length,

$x = \sqrt{3}$. Hence,

$y = \sqrt{\{2 - \sqrt{(4 - 3)}\}} = 1$(6 sides)

$z = \sqrt{\{2 - \sqrt{(4 - 1)}\}} = \sqrt{(2 - \sqrt{3})}$(12 sides)

...

$u = 0.065\,438\,52$ (96 sides)

Hence the polygon with 96 sides has a perimeter of

$$96 \times 0.065\,438\,52 = 6.282\,097\,9. \quad \text{So, } 2\pi = 6.282\,097\,9$$

and $$\pi = 3.141\,048\,9$$

3 The quantity b satisfies the equation, $b = 1/(1 + b)$ so,

$$b^2 + b - 1 = 0 \quad \text{i.e.} \quad b = (-1 \pm \sqrt{5})/2.$$

4 (i) Suppose that $p/q + e = \sqrt{2}$. (i.e. that the error in p/q as an approximation to $\sqrt{2}$ is e.) Then:

$$p^2/q^2 + 2p/q.e + e^2 = 2$$

If e is small enough, e^2 can be neglected so that,

$$e = \frac{2 - p^2/q^2}{2p/q} = \frac{2q^2 - p^2}{2pq}$$

and then

$$\sqrt{2} = p/q + e = p/q + \frac{2q^2 - p^2}{2pq} = \frac{p^2 + 2q^2}{2pq}$$

(ii) Let us start the process with $p/q = 1/1 = 1$ as an approximation to $\sqrt{2}$. Using the formula above, we obtain:

Approximation to $\sqrt{2}$	1	3/2	17/12	577/408 ...
Likely error	$\frac{1}{2}$	$-1/12$	1/408	...

5 Suppose that $f(a,b) = f(p,q)$ (i.e. that f is not one–one.)
Then, $$2^{a-1}(2b - 1) = 2^{p-1}(2q - 1). \text{ Hence}$$
$$2^{a-1}(2b - 1) - 2^{p-1}(2q - 1) = 0. \text{ Assume that } a > p,$$
$$2^{p-1}(2^{a-p}(2b - 1) - 2q + 1) = 0 \text{ Now } 2^{p-1} \neq 0$$

for any natural number p; hence,

$$2^{a-p}(2b - 1) - 2q + 1 = 0 \text{ and so } 2^{a-p}(2b - 1) = 2q - 1$$

In this last equation the l.h.s is an even number, while the r.h.s is not. Hence, $a - p = 0$ i.e. $a = p$ and so $b = q$.
(ii) Clearly from the last result, $\mathscr{C}(N \times N) = \aleph_0$. It is easy to extend this result

(by the same function) to show that,

$$\mathscr{C}(N \times N \times N \times \ldots \times N) = \aleph_0.$$

6 (i) If n is not prime, we may write $n = rs$. Then,

$$M_n = 2^{r \cdot s} - 1 = (2^r)^s - 1$$

Using the identity, $x^s - 1 = (x-1)(x^{s-1} + x^{s-2} + \ldots + 1)$ and putting, $x = 2^r$, obviously M_n is not prime.

(ii) $M_{11} = 2^{11} - 1 = 2047$

7 A number with 10 digits is such that $10^9 \leq N < 10^{10}$. Hence the number of primes less than 10^{10} is

$$= 10^{10}/\log_e 10^{10} = 434\,300\,000$$

Similarly, the number of primes less than 10^9 is

$$10^9/\log_e 10^9 = 48\,260\,000.$$

So the number of primes with ten digits is 386 000 000.

8 (i) Since $\sqrt{8453} = 91.94, 91^2 < 8453 < 92^2$.
Hence, we obtain the sequence (from $n = 92$, $N = 8453$):

$92^2 - 8453, 93^2 - 8453, 94^2 - 8453, \ldots$ which are,

11 , 196 , ... and we need go no further,

since $196 = 14^2$. Hence,

$$93^2 - 8453 = 14^2 \text{ and so } 8453 = 93^2 - 14^2$$
$$\text{i.e. } 8453 = (93 + 14)(93 - 14) = 107.79, \text{ and so.}$$

8453 is not prime.

(ii) In the sequence, we obtain; $n^2 - N, (n+1)^2 - N, \ldots$
If N were prime, the process goes on until we reach a point where

$$(n+r)^2 - N = y^2 \text{ whence } N = (n+r+y)(n+r-y)$$

in which $n + r + y = N$ and $n + r - y = 1$.

9 Suppose that p is an odd prime. Then $(p+1)/2$ and $(p-1)/2$ are both natural numbers. (Why ?) Moreover,

$$(p + 1/2)^2 - (p - 1/2)^2 = p$$

Further, suppose that $p = x^2 - y^2$ then $p = (x-y)(x+y)$. But p is prime and so $x - y = 1$ and $x + y = p$. Solving, we find

$$x = (p+1)/2 \quad \text{and} \quad y = (p-1)/2$$

Hence every prime can be expressed (uniquely) as the difference between the squares of two natural numbers.

10 From the last equation r_n must divide r_{n-1}. Applying this to the second last equation, r_n must also divide r_{n-2}. Running backwards through all these equations, and ending at the first,

$$a = q_1 b + r_1$$

this means that r_n divides a and b. Since the remainders $r_1, r_2, \ldots$ are the 'highest' at each stage, we conclude that the last non-zero remainder (r_n) is the greatest common divisor of a and b.

(ii) Using the last but one equation,

$$\begin{aligned}
d = r_n &= r_{n-2} - q_n r_{n-1} \\
&= r_{n-2} - q_n(r_{n-3} - q_{n-1}r_{n-2}) \\
&= -q_n r_{n-3} + (1 + q_n q_{n-1})r_{n-2} \\
&= \dots \\
&= (\quad)r_1 + (\quad)r_2 \\
&= (\quad)r_1 + (\quad)(b - q_2 r_1) \\
&= ((\quad) - q_2(\quad))r_1 + (\quad)b \\
&= ((\quad) - q_2(\quad))(a - q_1 b) + (\quad)b \\
&= (\text{integer})\, a + (\text{integer})\, b
\end{aligned}$$

11 Let us suppose that p does not divide a. since p is prime they cannot have any common factors other than 1. Hence,

$$1 = (p, a)$$

and using this last result we can find integers x and y so that,

$$xp + ya = 1$$
$$xpb + yab = b$$

The l.h.s. of this equation is divisible by p because xpb contains p, and yab contains ab which we are given is divisible by p. Hence the r.h.s. is also divisible by p. Hence b is divisible by p. (N.B. This result is crucial. For instance, when trying to prove on p. 39 that $\sqrt{5}$ is irrational we show that $5q^2 = p^2$. Hence p^2 is divisible by 5, and since 5 is prime, p is also divisible by 5.)

12 If 1 were a prime number, we could write

$$12 = 2^2.3.1 = 2^2.3.1.1$$

which would not give us unique factorization of the integers into primes.

Chapter 6

Q 6.2 Suppose that $ab \equiv ac \pmod{m}$ and that a and m are relatively prime. Then

$$ab = ac + pm, \text{ for some integer } p \tag{1}$$

If a and m are relatively prime they have no common divisors other than 1. Hence (Chapter 5 Further Investigations 10) we can find integers x and y so that

$$xa + ym = 1 \tag{2}$$

Multiplying (1) by y and (2) by p gives

$$aby = acy + pmy \text{ and } xap + ypm = p$$

hence, $$aby = acy + (p - xap)$$

i.e. $$a(by - cy + xp) = p$$

So p is divisible by a. We can write, $p = ra$, and putting this back into (1) we have

$$\begin{aligned}
&ab = ac + ram \\
\Rightarrow\ &b = c + rm \\
\Rightarrow\ &b \equiv c \pmod{m}
\end{aligned}$$

Q 6.3 Let N be any positive integer. Then in the decimal notation we may write

$$N = a + b.10 + c.10^2 + \dots$$

Now,

$$10 \equiv 1 \pmod 9 \Rightarrow b.10 \equiv b \pmod 9,$$
$$100 \equiv 1 \pmod 9 \Rightarrow c100 \equiv c \pmod 9, \text{etc.}$$

Then

$$N = a + b.10 + c.100 + \dots \equiv a + b + c + \dots$$

Hence N is divisible by 9 only if $a + b + c \dots$ is.

Q 6.4 (i) In mod 3 we have only the elements 0, 1, and 2. Let $x \equiv 3/4 \pmod 3$. Then $4x \equiv 3 \Rightarrow 4x \equiv 0 \pmod 3$ and so $x \equiv 0$. Hence $\frac{3}{4} \equiv 0 \pmod 3$.
(ii) Let $y \equiv 2/5 \pmod 3$, $5y \equiv 2$. However, $5.1 \equiv 2 \pmod 3$ and no other element satisfies this. Hence, $y \equiv 1$, and so $2/5 \equiv 1 \pmod 3$
(iii) Let $z \equiv 23/20 \pmod 3$. Then $z \equiv 1$.
(iv) Yes, because $0 + 1 = 1$.

Q 6.5 If $2x - 6y = 2n + 1$ then, $2x = 6y + 2n + 1$, $2x \equiv 2n + 1 \pmod 6$. Now $2n + 1$ is always odd. However, $2x \pmod 6$ can take any of the values; $2.0 \equiv 0, 2.1 \equiv 1, 2.2 \equiv 4$, $2.3 \equiv 0$, $2.4 \equiv 2$, or $2.5 \equiv 4 \pmod 6$. That is, only an even value. Hence, there can be no solution.

Q 6.6 If $x \equiv 3 \pmod 5$ then $x^2 \equiv 9 \equiv 4$, $x^3 \equiv 27 \equiv 2 \pmod 5$, $8x \equiv 24 \equiv 4$, and $6 \equiv 1 \pmod 5$. Hence $f(x) \equiv 2 - 4 + 1 \equiv -1 \equiv 4 \pmod 5$.

Q 6.7 The equation may be written $x^{2m} + x^{2m-1} + x^{2m-2} + \dots + 1 = 0$. Consider the function

$$f(x) = x^{2m} + x^{2m-1} + \dots + 1$$

Then in mod 2 we have $f(0) = 0 + 0 + \dots + 1 \equiv 1 \pmod 2$

and $$f(1) = 1 + 1 + 1 + 1 + \dots + 1 = 2m + 1 \equiv 1 \pmod 2.$$

Since $f(x) \not\equiv 0 \pmod 2$ for all x, the equation $f(x) = 0$ has no integer and hence no rational solutions.

Q 6.9 Let $g(x) = 8x^3 + 4x^2 - 4x - 1$. In mod 3 we have,

$x(\equiv)$	0	1	2
$f(x)(\equiv)$	2	1	2

Q 6.10

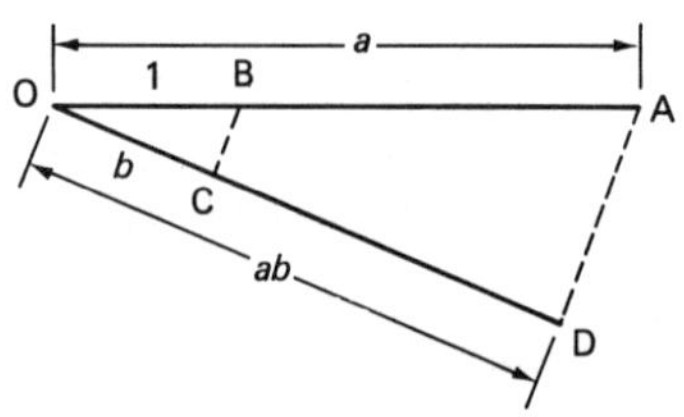

Q 6.12 If we can construct a nine-sided regular polygon, we should have constructed an angle of 40° (360°/9). By bisecting this we should obtain 20°, which is, of course, one third of 60°. Hence, we should be able to trisect 60°.

Q 6.13 From the equation $\cos 4\theta = \cos 3\theta$, one solution is $\theta = 0°$ which means that $\cos \theta = x = 1$. This is obviously not the solution we require.

Further investigations

1 If the polygon is drawn in a circle, bisect each of the angles made by two vertices with the centre. Joining up these new points on the circumference yields a regular polygon with $2m$ sides.

2 In constructing a regular pentagon we shall need an angle of $(360°/5) = 72°$. In other words we shall need to solve the equation

$$5\theta = 360° \Rightarrow 3\theta = (360 - 2\theta)° \Rightarrow \cos 3\theta = \cos(360 - 2\theta)$$
$$\text{i.e. } \cos 3\theta = \cos 2\theta$$

Hence, $4\cos^3\theta - 3\cos\theta = 2\cos^2\theta - 1$, which gives the equation

$$4x^3 - 2x^2 - 3x + 1 = 0 \; (x = \cos\theta)$$

This has a solution of $x = 1$ (why?) and when this is factored out (i.e. divide by $x - 1$) we get, $4x^2 + 2x - 1 = 0$. This has roots $x = \cos\theta = (-1 \pm \sqrt{5})/4$. Clearly the negative sign in front of the square root is inappropriate, and so $\cos 72° = (-1 + \sqrt{5})/4$. This length is easily constructed, and from a unit circle we can now obtain 72° as shown in the Figure.

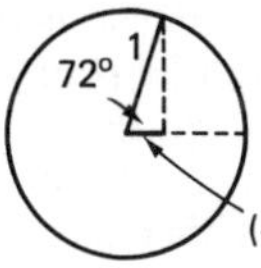

3 The three sided regular polygon yields an angle of 60°. From this we can make any multiple of 60°, i.e. 120°. From the regular pentagon we obtain an angle of 72°, and hence the angle $2.72° = 144°$. Now, $144° - 120° = 24°$, which is the angle required to draw a regular polygon of 15 sides $(360/15 = 24)$.

4 (i) $51 = 3 \times 17$ and 3 and 7 are relatively prime; hence their greatest common divisor is 1. Using the result of Chapter 5, Further Investigations question 10, there exist integers x and y so that

$$1 = x{\cdot}3 + y{\cdot}17$$

In fact, $x = 6$ and $y = -1$ so; $1 = 6{\cdot}3 + (-1){\cdot}17$ and hence,

$$1/51 = 6{\cdot}/17 - 1/3$$
$$\Rightarrow \; 360/51 = 6{\cdot}(360/17) - (360/3)$$

Hence from the three sided regular polygon and the 17 sided regular polygon we can construct an angle of $(360/51)°$ which will give us the 51 sided regular polygon.

(ii) $85 = 5.17$

6 Yes.

7 Question 4 was the lead in to this. Suppose that the greatest common divisor of p and q is d. Then we can find integers x and y so that $d = xp + yq$, hence $d/pq = x/q + y/p$, which gives, $d(360/pq) = x(360/q) + y(360/p)$
To draw a regular polygon of (pq) sides we need the angle $360/pq$. This last equation says we can get the dth multiple of it. Thus if $d = 1$ (i.e. p and q are relatively prime) we can construct the polygon. If d is a power of 2 we can bisect to obtain the required angle.

8 Obviously we want the value of 3^{89} in mod 10. Now $3 \equiv 3 \pmod{10}$
$3^2 = 9 \equiv -1 \pmod{10}, 3^4 \equiv 1, 3^8 \equiv 1, \ldots 3^{64} \equiv 1 \pmod{10}$.
But, $3^{89} = 3^{64+16+8+1} = 3^{64}.3^{16}.3^8.3^1 \equiv 1{\cdot}1{\cdot}1{\cdot}3 \pmod{10}$ Hence, $3^{89} \equiv 3 \pmod{10}$

9 This time we want the value of 3^{89} in mod 100. Proceeding in the same way: $3 \equiv 3, 3^2 \equiv 9, 3^4 \equiv 81 \equiv -19, 3^8 \equiv 361 \equiv 61 \equiv -39, 3^{16} \equiv 1521 \equiv 21, \ldots, 3^{64} \equiv 1681 \equiv 81 \equiv -19 \pmod{100}$.
Hence, $3^{89} \equiv (-19)\cdot(21)\cdot(-39)\cdot(3) \equiv 46\,683 \equiv 83 \pmod{100}$.

10 Let $P(n) = 2^{2^n} + 1$. Then, $P(n) - 1 = 2^{(2^n)}$.
Now, $\{2^n\} = \{2, 4, 8, 16, \ldots\}$ and hence,

$$\{P(n) - 1\} = \{2^2, 2^4, 2^8, 2^{16}, \ldots\}$$

But, $2 \equiv 2, 2^2 \equiv 4, 2^4 \equiv 16 \equiv 6, 2^8 \equiv 36 \equiv 6 \pmod{10}$, and so $2^{16}, 2^{32}, \ldots$ are all such that they are congruent to $6 \pmod{10}$. So $P(n) - 1 \equiv 6 \Rightarrow P(n) \equiv 7 \pmod{10}$ when $n > 1$ and so they end in 7.

Chapter 7

Q 7.1 The following are possibilities:
(2) It is possible to extend a controversy indefinitely.
(3) It is possible to construct a circular argument with any point of view as subject, and at any distance from that subject.
(4) All right arguments are equally worthy.
(5) If one controversy becomes involved with two others so as to make the difference between the first and either of the second controversies less than a right argument, the second two controversies will, if indefinitely extended, be resolved.

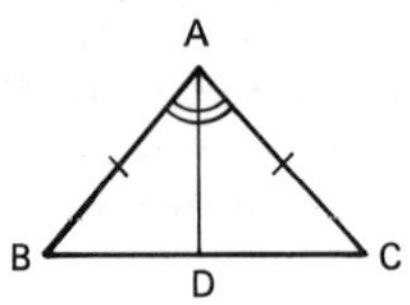

Q 7.2 Clearly $\triangle ABD \cong \triangle ADC$ because
AB = AC (given)
$\angle$ BAD = $\angle$ DAC (constr.)
AD = AD (identity)
Hence, we conclude that $\angle B = \angle C$ as required.

Q 7.3 Place the two triangles together as shown in the Figure. Now
AC = AF (given)
$\Rightarrow$ $\triangle$ AFC is isosceles.
$\Rightarrow$ $\angle x = \angle y$ (Theorem 2)
Similarly FB = BC (given), and so $\triangle$ FBC is isosceles $\Rightarrow \angle p = \angle r$ (Theorem 2). Hence, $\angle$ AFB = $\angle$ ACB and so $\triangle ABC \cong \triangle DFE$ (Theorem 1) as required.

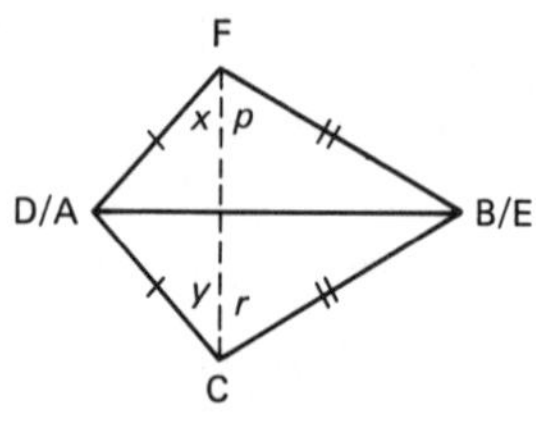

Q 7.4

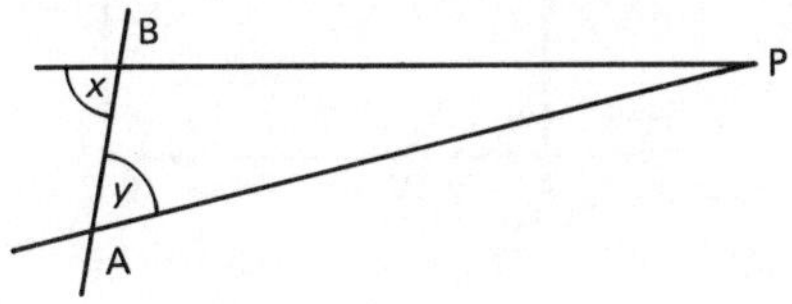

Suppose the lines meet at P. Consider $\triangle$ ABP. Now by Theorem 5, the exterior angle of a triangle is greater than either remote interior angles. So, $\angle x > \angle y$. But $\angle x = \angle y$ (given), which is a contradiction, and so the lines cannot meet.

Q 7.5 Suppose P is the point and h the line. Through P draw any line AB cutting h at B. At P construct, $\angle y = \angle x$ to give required line h′. Then by Theorem 6, this line is parallel to h.

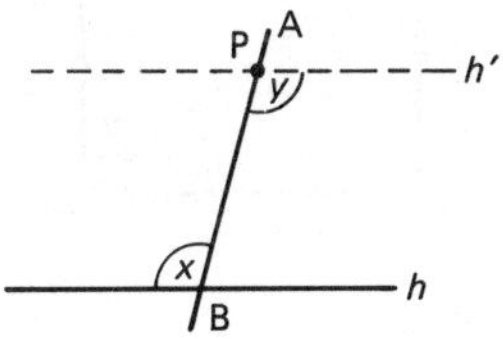

Q 7.6 Draw in the two diagonals of the quadrilateral as shown. Then $\triangle$ DAB $\cong$ $\triangle$ CBA because

CB = DA (given)
$\angle A = \angle B$ (given)
and AB is common (Theorem 1).

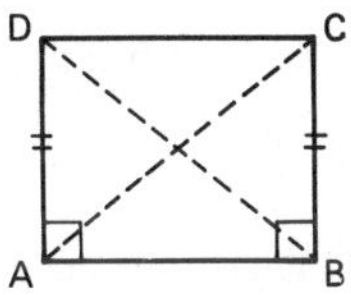

Hence, DB = AC, and $\angle$ CAB = $\angle$ DBA. This last statement means that, $\angle$ CAD = $\angle$ DBC (why?). Now, $\triangle$ ADC $\cong$ $\triangle$ BCD because, AD = BC (given), AC = DB (just proved), and $\angle$ CAD = $\angle$ DBC (again, just proved). So Theorem 1 applies again, and we conclude $\angle D = \angle C$ as required.

Q 7.7 If we accept the hypothesis of the right-angle, we have the figure shown. If the mid-points of AD and BC are respectively P and Q, we can place quadrilateral PQCD over the quadrilateral PQBA so that $\angle C$ coincides with $\angle B$. Then since BQ = QC and $\angle B = \angle C$ this end of the quadrilaterals coincide completely. D must lie on AB since $\angle C = \angle B$. Suppose D does not coincide with A, and that it lies either at D′ or D″ as shown. Neither of these positions is possible because for D″, the external angle

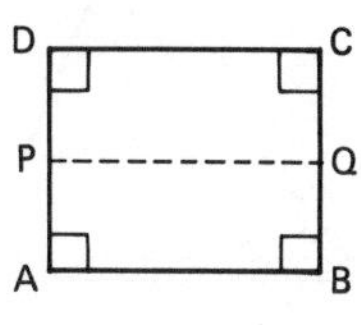

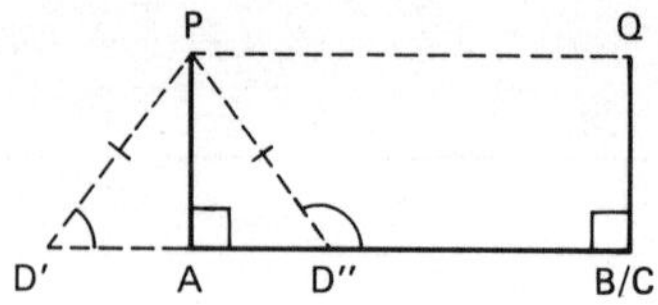

of $\triangle$ PAD (the right angle) is not greater than the opposite interior angle A (another right angle). This is also true for D′. Hence, D coincides with A and so, DC = AB. We now have the following figure.

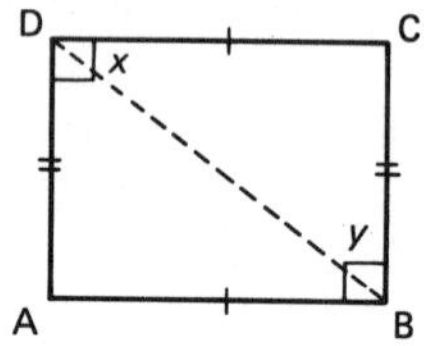

Now $\triangle$ ADB $\cong$ $\triangle$ CBD because,
DC = AB (just proved)
DA = CB (given)
and $\angle C = \angle A$ (given)
and Theorem 1. Hence, $\angle$ ABD = x, but $\angle$ ABD + y = a right angle so $x + y$ = a right angle. So the angles of $\triangle$ BDC are x, y, and a right angle, which makes an angle sum of two right angles. Hence, we have established the first form of the parallel postulate, given on p. 103.

Q 7.8 Re-write the proof of Theorem 7 with acute replaced by either right or obtuse angle. The required results follow.

Q 7.9 To ensure that P lies between B and C. Actually, this is not provable on the axioms that we have taken. However, in a more complete set of axioms that Euclid used rather than stated, this would have to be included.

Q 7.10 Now $x + y + z < 180^\circ$. Let,

$$180 - (x + y + z) = \Delta \qquad (1)$$

Then in the Figure

$$x + \gamma + \alpha < 180^\circ$$

Suppose that,

$$180 - (x + \gamma + \alpha) = \Delta_1 \qquad (2)$$

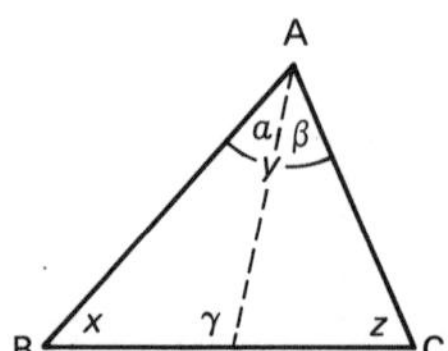

Again, $180 - \gamma + \beta + z < 180$, so that

$$180 - (180 - \gamma + \beta + z) = \Delta_2 \text{ say i.e.}$$

$$\gamma - \beta - z = \Delta_2 \qquad (3)$$

Adding (2) and (3)

$$180 - x - \gamma - \alpha + \gamma + \beta + z = \Delta_1 + \Delta_2$$

$$\Rightarrow \quad 180 - (x + y + z) = \Delta_1 + \Delta_2$$

using (1);

$$\Delta = \Delta_1 + \Delta_2$$

Q 7.11 Using the letters of the answer given to **Q 7.5**. Draw a line through the point P. How do we *know* that any line through P actually cuts the line *h*? Only if we assume the infinitude of this line can we guarantee that it meets the line *h*.

Q 7.12 'Suppose two spherical triangles are similar but not congruent. Since they are similar they have the same angle sum and hence the same excess. Hence their areas are the same. So we conclude that similar spherical triangles are also congruent.'

This proof is *not correct*. Where is the fallacy in it? The correct proof is the obvious modification of the proof of Theorem 9 in this chapter (p. 107). In order to correct the proof above, we should have to prove that if two triangles are not congruent their areas cannot be the same. This is quite a difficult proof.

Q 7.13 Yes. Consider two concentric spheres with a triangle drawn on the inner one. A 'similar' triangle on the outer sphere is produced by the projection lines emanating from the common centre, through the vertices of this triangle onto the outer sphere.

Q 7.14 Consider the triangle ABC. If we define angle in the usual way, then its angle sum is two right angles. However, in this geometry we know this not to be true. Hence we must find a different definition of angle.

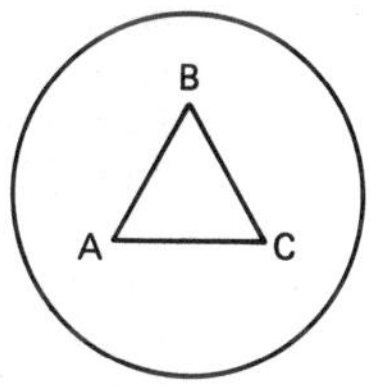

Q 7.15 (i) $\text{Distance (PQ)} = \log\dfrac{(QA)(PB)}{(PA)(QB)}$, $\quad \text{Distance (QP)} = \log\dfrac{(PB)(QA)}{(QB)(PA)}$

which are, of course, the same.

(ii) $\text{Distance (PQ)} = \log\dfrac{(QA)(PB)}{(PA)(QB)}$, $\quad \text{Distance (QR)} = \log\dfrac{(RA)(QB)}{(QA)(RB)}$

$$\text{Distance (PQ)} + \text{Distance (QR)} = \log\frac{(QA)(PB)}{(PA)(QB)} + \log\frac{(RA)(QB)}{(QA)(RB)}$$

$$= \log\frac{(QA)(PB)(RA)(QB)}{(PA)(QB)(QA)(RB)}$$

$$= \log\frac{(RA)(PB)}{(PA)(RB)}$$

$$= \text{Distance (PR)}$$

(iii) As Q moves towards B, $QA \to AB$ and $QB \to 0$. Hence, Distance $(PQ) \to \infty$.

Index